A VLSI ARCHITECTURE FOR CONCURRENT DATA STRUCTURES

THE KLUWER INTERNATIONAL SERIES
IN ENGINEERING AND COMPUTER SCIENCE

VLSI, COMPUTER ARCHITECTURE AND
DIGITAL SIGNAL PROCESSING

Consulting Editor

Jonathan Allen

Other books in the series:

Logic Minimization Algorithms for VLSI Synthesis, R.K. Brayton, G.D. Hachtel, C.T. McMullen, and A.L. Sangiovanni-Vincentelli. ISBN 0-89838-164-9.

Adaptive Filters: Structures, Algorithms, and Applications, M.L. Honig and D.G. Messerschmitt. ISBN: 0-89838-163-0.

Computer-Aided Design and VLSI Device Development, K.M. Cham, S.-Y. Oh, D. Chin and J.L. Moll. ISBN 0-89838-204-1.

Introduction to VLSI Silicon Devices: Physics, Technology and Characterization, B. El-Kareh and R.J. Bombard. ISBN 0-89838-210-6.

Latchup in CMOS Technology: The Problem and Its Cure, R.R. Troutman. ISBN 0-89838-215-7.

Digital CMOS Circuit Design, M. Annaratone. ISBN 0-89838-224-6.

The Bounding Approach to VLSI Circuit Simulation, C.A. Zukowski. ISBN 0-89838-176-2.

Multi-Level Simulation for VLSI Design, D.D. Hill, D.R. Coelho. ISBN 0-89838-184-3.

Relaxation Techniques for the Simulation of VLSI Circuits, J. White and A. Sangiovanni-Vincentelli. ISBN 0-89838-186-X.

VLSI CAD Tools and Applications, W. Fichtner and M. Morf. ISBN 0-89838-193-2.

A VLSI ARCHITECTURE FOR CONCURRENT DATA STRUCTURES

by

William J. Dally
Massachusetts Institute of Technology

KLUWER ACADEMIC PUBLISHERS
Boston/Dordrecht/Lancaster

Distributors for North America:
Kluwer Academic Publishers
101 Philip Drive
Assinippi Park
Norwell, Massachusetts 02061, USA

Distributors for the UK and Ireland:
Kluwer Academic Publishers
MTP Press Limited
Falcon House, Queen Square
Lancaster LA1 1RN, UNITED KINGDOM

Distributors for all other countries:
Kluwer Academic Publishers Group
Distribution Centre
Post Office Box 322
3300 AH Dordrecht, THE NETHERLANDS

Library of Congress Cataloging-in-Publication Data

Dally, William J.
  A VLSI architecture for concurrent data
structures.

  (The Kluwer international series in engineering
and computer science ; SECS 027)
  Abstract of thesis (Ph. D.)—California Institute
of Technology.
  Bibliography: p.
  1. Electronic digital computers—Circuits.
2. Integrated circuits—Very large scale integration.
3. Computer architecture. I. Title. II. Series.
TK7888.4.D34 1987   621.395   87–3350
ISBN 0–89838–235–1

Contents

List of Figures

Preface

Concurrent data structures simplify the development of concurrent programs by encapsulating commonly used mechanisms for synchronization and communication into data structures. This thesis develops a notation for describing concurrent data structures, presents examples of concurrent data structures, and describes an architecture to support concurrent data structures.

Concurrent Smalltalk (CST), a derivative of Smalltalk-80 with extensions for concurrency, is developed to describe concurrent data structures. CST allows the programmer to specify objects that are distributed over the nodes of a concurrent computer. These distributed objects have many *constituent objects* and thus can process many messages simultaneously. They are the foundation upon which concurrent data structures are built.

The *balanced cube* is a concurrent data structure for ordered sets. The set is distributed by a balanced recursive partition that maps to the subcubes of a binary n-cube using a Gray code. A search algorithm, VW search, based on the distance properties of the Gray code, searches a balanced cube in $O(\log N)$ time. Because it does not have the root bottleneck that limits all tree-based data structures to $O(1)$ concurrency, the balanced cube achieves $O(\frac{N}{\log N})$ concurrency.

Considering graphs as concurrent data structures, graph algorithms are presented for the shortest path problem, the max-flow problem, and graph partitioning. These algorithms introduce new synchronization techniques to achieve better performance than existing algorithms.

A message-passing, concurrent architecture is developed that exploits the characteristics of VLSI technology to support concurrent data structures. Interconnection topologies are compared on the basis of dimension. It is shown that minimum latency is achieved with a very low dimensional network. A deadlock-free routing strategy is developed for this class of networks, and a prototype VLSI chip implementing this strategy is described. A message-driven processor complements the network by responding to messages with a very low latency. The processor directly executes messages, eliminating a level of interpretation. To take advantage of the performance offered by specialization while at the same time retaining flexibility, processing elements can be specialized to operate on a single class of objects. These *object experts* accelerate the performance of all applications using this class.

This book is based on my Ph.D. thesis, submitted on March 3, 1986, and awarded the Clauser prize for the most original Caltech Ph.D. thesis in 1986. New material, based on work I have done since arriving at MIT in July of 1986, has been added to Chapter 5. The book in its current form presents a coherent view of the art of designing and programming concurrent computers. It can serve as a handbook for those working in the field, or as supplemental reading for graduate courses on parallel algorithms or computer architecture.

Acknowledgments

While a graduate student at Caltech I have been fortunate to have the opportunity to work with three exceptional people: Chuck Seitz, Jim Kajiya, and Randy Bryant. My ideas about the architecture of VLSI systems have been guided by my thesis advisor, Chuck Seitz, who also deserves thanks for teaching me to be less an engineer and more a scientist. Many of my ideas on object-oriented programming come from my work with Jim Kajiya, and my work with Randy Bryant was a starting point for my research on algorithms.

I thank all the members of my reading committee: Randy Bryant, Dick Feynman, Jim Kajiya, Alain Martin, Bob McEliece, Jerry Pine, and Chuck Seitz for their helpful comments and constructive criticism.

My fellow students, Bill Athas, Ricky Mosteller, Mike Newton, Fritz Nordby, Don Speck, Craig Steele, Brian Von Herzen, and Dan Whelan have provided constructive criticism, comments, and assistance.

This manuscript was prepared using TEX [75] and the LaTEX macro package [80]. I thank Calvin Jackson, Caltech's TEXpert, for his help with typesetting problems. Most of the figures in this thesis were prepared using software developed by Wen-King Su. Bill Athas, Sharon Dally, John Tanner, and Doug Whiting deserve thanks for their careful proofreading of this document.

Mike Newton of Caltech and Carol Roberts of MIT have been instrumental in converting this thesis into a book.

Financial support for this research was provided by the Defense Advanced Research Projects Agency. I am grateful to AT&T Bell Laboratories for the support of an AT&T Ph.D. fellowship.

Most of all, I thank Sharon Dally for her support and encouragement of my graduate work, without which this thesis would not have been written.

A VLSI ARCHITECTURE FOR
CONCURRENT DATA STRUCTURES

Chapter 1

Introduction

Computing systems have two major problems: they are too slow, and they are too hard to program.

Very large scale integration (VLSI) [88] technology holds the promise of improving computer performance. VLSI has been used to make computers less expensive by shrinking a rack of equipment several meters on a side down to a single chip a few millimeters on a side. VLSI technology has also been applied to increase the memory capacity of computers. This is possible because memory is incrementally extensible; one simply plugs in more chips to get a larger memory. Unfortunately, it is not clear how to apply VLSI to make computer systems faster. To apply the high density of VLSI to improving the speed of computer systems, a technique is required to make processors incrementally extensible so one can increase the processing power of a system by simply plugging in more chips.

Ensemble machines [112] , collections of processing nodes connected by a communications network, offer a solution to the problem of building extensible computers. These concurrent computers are extended by adding processing nodes and communication channels. While it is easy to extend the hardware of an ensemble machine, it is more difficult to extend its performance in solving a particular problem. The communication and synchronization problems involved in coordinating the activity of the many processing nodes make programming an ensemble machine difficult. If the processing nodes are too tightly synchronized, most of the nodes will remain idle; if they are too loosely synchronized, too much redundant work is performed. Because of the difficulty of programming an ensemble machine, most successful applications of these machines have been to problems where the structure of the data is quite regular, resulting in a regular communication pattern.

Object-oriented programming languages make programming easier by providing data abstraction, inheritance, and late binding [123]. Data abstraction separates an object's protocol, the things it knows how to do, from an object's implementation, how it does them. This separation encourages programmers to write modular code. Each module describes a particular type or *class* of object. Inheritance allows a programmer to define a *subclass* of an existing class by specifying only the differences between the two classes. The subclass inherits the remaining protocol and behavior from its *superclass*, the existing class. Late, run-time, binding of meaning to objects makes for more flexible code by allowing the same code to be applied to many different classes of objects. Late binding and inheritance make for very general code. If the problems of programming an ensemble machine could be solved inside a class definition, then applications could share this class definition rather than have to repeatedly solve the same problems, once for each application.

This thesis addresses the problem of building and programming extensible computer systems by observing that most computer applications are built around data structures. These applications can be made concurrent by using *concurrent data structures,* data structures capable of performing many operations simultaneously. The details of communication and synchronization are encapsulated inside the class definition for a concurrent data structure. The use of concurrent data structures relieves the programmer of many of the burdens associated with developing a concurrent application. In many cases communication and synchronization are handled entirely by the concurrent data structure and no extra effort is required to make the application concurrent. This thesis develops a computer architecture for concurrent data structures.

1.1   Original Results

The following results are the major original contributions of this thesis:

- In Section 2.2, I introduce the concept of a *distributed object*, a single object that is distributed across the nodes of a concurrent computer. Distributed objects can perform many operations simultaneously. They are the foundation upon which concurrent data structures are built.

- A new data structure for ordered sets, the *balanced cube*, is developed in Chapter 3. The balanced cube achieves greater concurrency than conventional tree-based data structures.

- In Section 4.2, a new concurrent algorithm for the shortest path problem is described.

- Two new concurrent algorithms for the max-flow problem are presented in Section 4.3.

- A new concurrent algorithm for graph partitioning is developed in Section 4.4.

- In Section 5.3.1, I compare the latency of k-ary n-cube networks as a function of dimension and derive the surprising result that, holding wiring bisection width constant, minimum latency is achieved at a very low dimension.

- In Section 5.3.2, I develop the concept of *virtual channels*. Virtual channels can be used to generate a deadlock-free routing algorithm for any *strongly connected* interconnection network. This method is used to generate a deadlock-free routing algorithm for k-ary n-cubes.

- The torus routing chip (TRC) has been designed to demonstrate the feasibility of constructing low-latency interconnection networks using *wormhole routing* and *virtual channels*. The design and testing of this self-timed VLSI chip are described in Section 5.3.3.

- In Section 5.5, I introduce the concept of an *object expert*, hardware specialized to accelerate operations on one class of object. Object experts provide performance comparable to that of special-purpose hardware while retaining the flexibility of a general purpose processor.

1.2 Motivation

Two forces motivate the development of new computer architectures: need and technology. As computer applications change, users need new architectures to support their new programming styles and methods. Applications today deal frequently with non-numeric data such as strings, relations, sets, and symbols. In implementing these applications, programmers are moving towards fine-grain object-oriented languages such as Smalltalk, where non-numeric data can be packaged into objects on which specific operations are defined. This packaging allows a single implementation of a popular object such as an ordered set to be used in many applications. These languages require a processor that can perform late binding of types and that can quickly allocate and de-allocate resources.

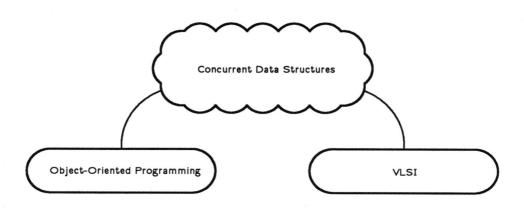

Figure 1.1: Motivation for Concurrent Data Structures

New architectures are also developed to take advantage of new technology. The emerging VLSI technology has the potential to build chips with 10^7 transistors with switching times of 10^{-10} seconds. Wafer-scale systems may contain as many as 10^9 devices. This technology is limited by its wiring density and communication speed. The delay in traversing a single chip may be 100 times the switching time. Also, wiring is limited to a few planar layers, resulting in a low communications bandwidth. Thus, architectures that use this technology must emphasize locality. The memory that stores data must be kept close to the logic that operates on the data. VLSI also favors specialization. Because a special purpose chip has a fixed communication pattern, it makes more effective use of limited communication resources than does a general purpose chip. Another way to view VLSI technology is that it has high throughput (because of the fast switching times) and high latency (because of the slow communications). To harness the high throughput of this technology requires architectures that distribute computation in a loosely coupled manner so that the latency of communication does not become a bottleneck.

This thesis develops a computer architecture that efficiently supports object-oriented programming using VLSI technology. As shown in Figure 1.1, the central idea of this thesis is concurrent data structures. The development of concurrent data structures is motivated by two underlying concepts: object-oriented programming and VLSI. The paradigm of object-oriented programming allows programs to be constructed from object classes that can be shared among applications. By defining concurrent data structures as distributed objects, these data structures can be shared across many applications. VLSI circuit technology motivates the use of concurrency and the construction of ensemble machines. These highly concurrent machines are required to take advantage of this high throughput, high latency technology.

1.3   Background

Much work has been done on developing data structures that permit concurrent access [33], [34], [35], [36], [78], [83]. A related area of work is the development of distributed data structures [41]. These data structures, however, are primarily intended for allowing concurrent access for multiple processes running on a sequential computer or for a data structure distributed across a loosely coupled network of computers. The concurrency achieved in these data structures is limited, and their analysis for the most part ignores communication cost. In contrast, the concurrent data structures developed here are intended for tightly coupled concurrent computers with thousands of processors. Their concurrency scales with the size of the problem, and they are designed to minimize communications.

Many algorithms have been developed for concurrent computers [7], [9], [15], [77] [87],[104], [118]. Most concurrent algorithms are for numerical problems. These algorithms tend to be oriented toward a small number of processors and use a MIMD [44] shared-memory model that ignores communication cost and imposes global synchronization.

Object-oriented programming began with the development of SIMULA [11], [19]. SIMULA incorporated data abstraction with classes, inheritance with subclasses, and late-binding with virtual procedures. SIMULA is even a concurrent language in the sense that it provides co-routining to give the illusion of simultaneous execution for simulation problems. Smalltalk [53], [54], [76], [138] combines object-oriented programming with an interactive programming environment. Actor languages [1], [17] are concurrent object-oriented languages where objects may send many messages without waiting for a reply. The programming notation used in this thesis combines the syntax of Smalltalk-80 with the semantics of actor languages.

The approach taken here is similar in many ways to that of Lang [81]. Lang also proposes a concurrent extension of an object-oriented programming language,

SIMULA, and analyzes communication networks for a concurrent computer to support this language. There are several differences between Lang's work and this thesis. First, this work develops several programming language features not found in Lang's concurrent SIMULA: distributed objects to allow concurrent access, simultaneous execution of several methods by the same object, and locks for concurrency control. Second, by analyzing interconnection networks using a wire cost model, I derive the result that low dimensional networks are preferable for constructing concurrent computers, contradicting Lang's result that high dimensional binary n-cube networks are preferable.

1.4   Concurrent Computers

This thesis is concerned with the design of concurrent computers to manipulate data structures. We will limit our attention to *message-passing* [114] MIMD [44] concurrent computers. By combining a processor and memory in each node of the machine, this class of machines allows us to manipulate data locally. By using a *direct* network, message-passing machines allow us to exploit locality in the communication between nodes as well.

Concurrent computers have evolved out of the ideas developed for programming multiprogrammed, sequential computers. Since multiple processes on a sequential computer communicate through shared memory, the first concurrent computers were built with shared memory. As the number of processors in a computer increased, it became necessary to separate the communication channels used for communication from those used to access memory. The result of this separation is the message-passing concurrent computer.

Concurrent programming models have evolved along with the machines. The problem of synchronizing concurrent processes was first investigated in the context of multiple processes on a sequential computer. This model was used almost without change on shared-memory machines. On message-passing machines, explicit communication primitives have been added to the process model.

1.4.1   Sequential Computers

A sequential computer consists of a processor connected to a memory by a communication channel. As shown in Figure 1.2, to modify a single data object requires three messages: an address message from processor to memory, a data message back to the processor containing the original object, and a data message

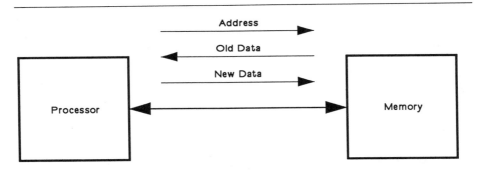

Figure 1.2: Information Flow in a Sequential Computer

back to memory containing the modified object. The single communication channel over which these messages travel is the principal limitation on the speed of the computation, and has been referred to as the Von Neumann bottleneck [4].

Even when a programmer has only a single processor, it is often convenient to organize a program into many concurrent processes. *Multiprogramming* systems are constructed on sequential computers by multiplexing many *processes* on the single processor. Processes in a multiprogramming system communicate through shared memory locations. Higher level communication and synchronization mechanisms such as interlocked read-modify-write operations, semaphores, and critical sections are built up from reading and writing shared memory locations. On some machines interlocked read-modify-write operations are provided in hardware.

Communication between processes can be synchronous or asynchronous. In programming systems such as CSP [64] and OCCAM [66] that use synchronous communication, the sending and receiving processes must *rendezvous*. Whichever process performs the communication action first must wait for the other process. In systems such as the Cosmic Cube [125] and actor languages [1],[17] that use asynchronous communication, the sending process may transmit the data and then proceed with its computation without waiting for the receiving process to accept the data.

Since there is only a single processor on a sequential computer, there is a unique global ordering of communication events. Communication also takes place without delay. A shared memory location written by process A on one memory cycle can be read by process B on the next cycle [1]. With global ordering of events and instantaneous communication, the strong synchronization implied by synchronous communication can be implemented without significant cost. The same is not true of concurrent computers where communication events are not uniquely ordered and the delay of communication is the major cost of computation.

It is possible for concurrent processes on a sequential computer to access an object *simultaneously* because the access is not really simultaneous. The processes, in fact, access the object one at a time. On a concurrent computer the illusion of simultaneous access can no longer be maintained. Most memories have a single port and can service only a single access at a given time.

1.4.2   Shared-Memory Concurrent Computers

To eliminate the Von Neumann bottleneck, the processor and memory can be replicated and interconnected by a switch. Shared memory concurrent computers such as the NYU Ultracomputer [108],[56],[57], C.MMP [137], and RP3 [102] consist of a number of processors connected to a number of memories through a switch, as shown in Figure 1.3.

Although there are many paths through the switch, and many messages can be transmitted simultaneously, the switch is still a bottleneck. While the bottleneck has been made wider, it has also been made longer. Every message must travel from one side of the switch to the other, a considerable distance that grows larger as the number of processors increases. Most shared-memory concurrent computers are constructed using *indirect* networks and cannot take advantage of locality. All messages travel the same distance regardless of their destination.

Shared-memory computers are programmed using the same process-based model of computation described above for multiprogrammed sequential computers. As the name implies, communication takes place through shared memory locations. Unlike sequential computers, however, there is no unique global order of communication events in a shared-memory concurrent computer, and several processors cannot access the same memory location at the same time.

Some designers have avoided the uniformly high communication costs of shared-memory computers by placing cache memories in the processing nodes [55].

---

[1]Some sequential computers overlap memory cycles and require a delay to read a location just written.

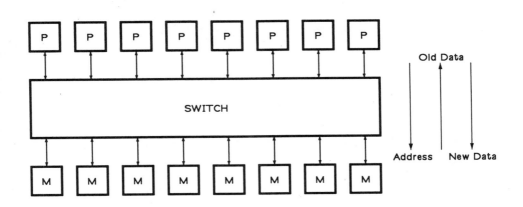

Figure 1.3: Information Flow in a Shared-Memory Concurrent Computer

Using a cache, memory locations used by only a single processor can be accessed without communication overhead. Shared memory locations, however, still require communication to synchronize the caches[2]. The cache nests the communication channel used to access local memory inside the channel used for interprocessor communication. This division of function between memory access and communication is made more explicit in message-passing concurrent computers.

1.4.3 Message-Passing Concurrent Computers

In contrast to sequential computers and shared-memory concurrent computers which operate by sending messages between processors and memories, a message-passing concurrent computer operates by sending messages between processing nodes that contain both logic and memory.

As shown in Figure 1.4, message-passing concurrent computers such as the Caltech Cosmic Cube [114] and the Intel iPSC [67] consist of a number of processing nodes interconnected by communication channels. Each processing node contains both a processor and a local memory. The communication channels used

---

[2]The problem of synchronizing cache memories in a concurrent computer is known as the cache *coherency* problem.

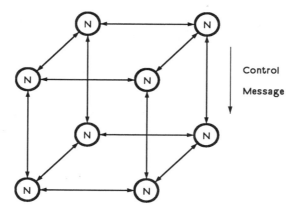

Control

Message

Figure 1.4: Information Flow in a Message-Passing Concurrent Computer

for memory access are completely separate from those used for inter-processor communication.

Message-passing computers take a further step toward reducing the Von Neumann bottleneck by using a *direct* network which allows locality to be exploited. A message to an object resident in a neighboring processor travels a variable distance which can be made short by appropriate process placement.

Shared-memory computers, even implemented with direct networks, use the available communications bandwidth inefficiently. Three messages are required for each data operation. A message-passing computer can make more efficient use of the available communications bandwidth by keeping the data state stationary and passing control messages. Since a processor is available at every node, data operations are performed in place. Only a single message is required to modify a data object. The single message specifies: the object to be modified, the modification to be performed, and the location to which the control state is to move next.

Keeping data stationary also encourages locality. Each data object is associated with the procedures that operate on it. This association allows us to place the logic that operates on a class of objects in close proximity to the memory that stores instances of the objects. As Seitz points out, "both the cost and performance metrics of VLSI favor architectures in which communication is localized" [113].

Message-passing concurrent computers are programmed using an extension of the process model that makes communication actions explicit. Under the Cosmic Kernel [125], for example, a process can send and receive messages as well as spawn other processes. This model makes the separation of communication from memory visible to the programmer. It also provides a base upon which an object-oriented model of computation can be built.

1.5 Summary

In this thesis I develop an architecture for concurrent data structures. I begin in Chapter 2 by developing the concept of a distributed object. A programming notation, Concurrent Smalltalk (CST), is presented that incorporates distributed objects, concurrent execution and locks for concurrency control. In Chapter 3 I use this programming notation to describe the balanced cube, a concurrent data structure for ordered sets. Considering graphs as concurrent data structures, I develop a number of concurrent graph algorithms in Chapter 4. New algorithms are presented for the shortest path problem, the max-flow problem, and graph partitioning. Chapter 5 develops an architecture based on the properties of the algorithms developed in Chapters 3 and 4 and the characteristics of VLSI technology. Network topologies are compared on the basis of dimension, and it is shown that low dimensional networks give lower latency than high dimensional networks for constant wire cost. A new algorithm is developed for deadlock-free routing in k-ary n-cube networks, and a VLSI chip implementing this algorithm is described. Chapter 5 also outlines the architecture of a message driven processor and describes how object experts can be used to accelerate operations on common data types.

Chapter 2

Concurrent Smalltalk

The message-passing paradigm of object-oriented languages such as Smalltalk-80 [53] introduces a discipline into the use of the communication mechanism of message-passing concurrent computers. Object-oriented languages also promote locality by grouping together data objects with the operations that are performed on them.

Programs in this thesis are described using Concurrent Smalltalk (CST), a derivative of Smalltalk-80 with three extensions. First, messages can be sent concurrently without waiting for a reply. Second, several methods may access an object concurrently. Locks are provided for concurrency control. Finally, the language allows the programmer to specify objects that are distributed over the nodes of a concurrent computer. These distributed objects have many *constituent objects* and thus can process many messages simultaneously. They are the foundation upon which concurrent data structures are built.

The remainder of this chapter describes the novel features of Concurrent Smalltalk. This discussion assumes that the reader is familiar with Smalltalk-80 [53]. A brief overview of CST is presented in Appendix A. In Section 2.1 I discuss the object-oriented model of programming and show how an object-oriented system can be built on top of the conventional process model. Section 2.2 introduces the concept of distributed objects. A distributed object can process many requests simultaneously. Section 2.3 describes how a method can exploit concurrency in processing a single request by sending a message without waiting for a reply. The use of *locks* to control simultaneous access to a CST object is described in Section 2.4. Section 2.5 describes how CST blocks include local variables and locks to permit concurrent execution of a block by the members of a collection. This chapter concludes with a brief discussion of performance metrics in Section 2.6.

2.1   Object-Oriented Programming

Object-oriented languages such as SIMULA [11] and Smalltalk [53] provide data abstraction by defining *classes* of objects. A class specifies both the data state of an object and the procedures or methods that manipulate this data.

Object-oriented languages are well suited to programming message-passing concurrent computers for four reasons.

- The message-passing paradigm of languages like Smalltalk introduces a discipline into the use of the communication mechanism of message-passing computers.

- These languages encourage locality by associating each data object with the methods that operate on the object.

- The information hiding provided by object-oriented languages makes it very convenient to move commonly used methods or classes into hardware while retaining compatibility with software implementations.

- Object names provide a uniform address space independent of the physical placement of objects. This avoids the problems associated with the partitioned address space of the process model: memory addresses internal to the process and process identifiers external to the process. Even when memory is shared, there is still a partition between memory addresses and process identifiers.

In an object-oriented language, computation is performed by sending messages to objects. Objects never wait for or explicitly receive messages. Instead, objects are reactive. The arrival of a message at an object triggers an *action*. The action may involve modifying the state of the object, transmitting messages that continue the control flow, and/or creating new objects.

The behavior of an object can be thought of as a function, B [1]. Let S be the set of all object states and M the set of all messages. An object with initial state, $i \in S$, receiving a message, $m \in M$, transitions to a new state, $n \in S$, transmits a possibly empty set of messages $m' \subset M$, and creates a possibly empty set of new objects $o \subset O$.

$$B : S \times M \to P(M), S, P(O). \tag{2.1}$$

Actions as described by the behavior function (2.1) are the primitives from which more complex computations are built. In analyzing timing and synchronization each action is considered to take place instantaneously, so it is possible to totally order the actions for a single object.

Methods are constructed from a set of primitive *actions* by sequencing the actions with messages. Often a method will send a message to an object and wait for a reply before proceeding with the computation. For example, in the code fragment below, the message size is sent to object x, and the method must wait for the reply before continuing.

```
xSize ←x size.
ySize ←xSize * 2.
```

Since there is no receive statement, multiple actions are required to implement this method. The first action creates a *context* and sends the size message. The context contains all method state: a pointer to the receiver, temporary variables, and an instruction pointer into the method code. A pointer to the context is placed in the reply-to field of the size message to cause the size method to reply to the context rather than to the original object. When the size method replies to the context, the second action resumes execution by storing the value of the reply into the variable xSize. The context is used to hold the state of the method between actions.

Objects with behaviors specified by (2.1) can be constructed using the message-passing process model. Each object is implemented by a process that executes an endless receive-dispatch-execute loop. The process receives the next message, dispatches control to the associated action, and then executes the action. The action may change the state of the object, send new messages, and/or create new objects. In Chapter 5 we will see how, by tailoring the hardware to the object model, we can make the receive-dispatch-execute process very fast.

2.2 Distributed Objects

In many cases we want an object that can process many messages simultaneously. Since the actions on an object are ordered, simultaneous processing of messages is not consistent with the model of computation described above. We can circumvent this limitation by using a distributed object. A distributed object consists of a collection of *constituent objects*, each of which can receive messages on behalf of the distributed object. Since many constituent objects

---

class	TallyCollection	*the class name*
superclass	Distributed Collection	*a distributed object*
instance variables	data	*local collection of data*
class variables		*none*
locks		*none*
class methods		

    class methods ...

instance methods

    tally: aKey   *count data matching aKey*
      | |
      (self upperNeighbor) localTally: aKey sum: 0 returnFrom: myId

    localTally: aKey sum: anInt returnFrom: anId
      | newSum |
      newSum ←anInt.
      data do: [:each |
        (each = aKey) ifTrue: [newSum ←newSum +1]].
      (myId = anId) ifTrue: [requester reply: newSum]
      ifFalse: [(self upperNeighbor) localTally: aKey sum: newSum returnFrom: anId].

    other instance methods ...

Figure 2.1: Distributed Object Class Tally Collection

---

can receive messages at the same time, the distributed object can process many messages simultaneously.

Figure 2.1 shows an example CST class definition. The definition begins with a header that identifies the name of the class, **Tally Collection**, the superclass from which **Tally Collection** inherits behavior, Distributed Collection, and the instance variables and locks that make up the state of each instance of the class. The header is followed by definitions of class methods, omitted here, and definitions of instance methods. Class methods define the behavior of the class object, **Tally Collection**, and perform tasks such as creating new instances of the class. Instance methods define the behavior of instances of class **Tally Collection**, the collections themselves. In Figure 2.1 two instance methods are defined.

Instances of class Tally Collection are distributed objects made up of many *constituent objects* (COs). Each CO has an instance variable data and understands the messages tally: and localTally:. A distributed object is created by sending a newOn message to its class.

a TallyCollection ←TallyCollection newOn: someNodes.

The argument of the newOn: message, someNodes, is a collection of processing nodes[1]. The newOn: message creates a CO on each member of someNodes. There is no guarantee that the COs will remain on these processing nodes, however, since objects are free to migrate from node to node.

When an object sends a message to a distributed object, the message may be delivered to any constituent of the distributed object. The sender has no control over which CO receives the message. The constituents themselves, however, can send messages to specific COs by using the message co:. For example, in the code below, the receiver (self), a constituent of a distributed object, sends a localTally message to the anId[th] constituent of the same distributed object.

(self co: anId) localTally: #foo sum: 0 returnFrom: myId.

The argument of the co: message is a *constituent identifier*. Constituent identifiers are integers assigned to each constituent sequentially beginning with one. The constant myId gives each CO its own index and the constant maxId gives each CO the number of constituents.

The method tally: aKey in Figure 2.1 counts the occurrences of aKey in the distributed collection and returns this number to the sender. The constituent object that receives the tally message sends a localTally message to its neighbor[2]. The localTally method counts the number of occurrences of aKey in the receiver node, adds this number to the sum argument of the message and propagates the message to the next CO. When the localTally message has visited every CO and arrives back at the original receiver, the total sum is returned to the original customer by sending a reply: message to requester.

Distributed objects often forward messages between COs before replying to the original requesting object. TallyCollection, for example, forwards localTally messages in a cycle to all COs before replying. CST supports this style of

---

[1] Processing nodes are objects.

[2] The message upperNeighbor returns the CO with identifier myId + 1 if myId $\neq$ maxId and the CO with identifier 1 otherwise.

programming by providing the reserved word requester. For messages arriving from outside the object, requester is bound to the sender. For internal messages, requester is inherited from the sending method.

This forwarding behavior illustrates a major difference between CST and Smalltalk-80: CST methods do not necessarily return a value to the sender. Methods that do not explicitly return a value using '↑' terminate without sending a reply. The tally: method terminates without sending a reply to the sender. The reply is sent later by the localTally method.

The tally: method shown in Figure 2.1 exhibits no concurrency. The point of a distributed object is not only to provide concurrency in performing a single operation on the object, but also to allow many operations to be performed concurrently. For example, suppose we had a Tally Collection with 100 COs. This object could receive 100 messages simultaneously, one at each CO. After passing 10,000 localTally messages internally, 100 replies would be sent to the original senders. The 100 requests are processed concurrently.

Some concurrent applications require global communication. For example, the concurrent garbage collector described by Lang [81] requires that processes running in each processor be globally synchronized. The hardware of some concurrent computers supports this type of global communication. The Caltech Cosmic Cube, for instance, provides several *wire-or* global communication lines for this purpose [114].

Some applications require global communication combined with a simple computation. For example, branch and bound search problems require that the minimum bound be broadcast to all processors. Ideally, a communication network would accept a bound from each processor, compute the minimum, and broadcast it. In fact, the computation can be carried out in a distributed manner on the *wire-or* lines provided by the Cosmic Cube.

Distributed objects provide a convenient and machine-independent means of describing a broad class of global communication services. The service is formulated as a distributed object that responds to a number of messages. For example, the synchronization service can be defined as an object of class Sync that responds to the message wait. The distributed object waits for a specified number of wait messages and then replies to all requesters. On machines that provide special hardware, class Sync can make use of this hardware. On other machines, the service can be implemented by passing messages among the constituent objects.

---

instance methods for class TallyCollection

tally: aKey *count data matching aKey*
| |
↑self localTally: aKey level: 0 root: myId.

localTally: aKey level: anInt root: anId
| upperTally lowerTally sum aLevel|
aLevel = anInt + 1.
sum ←0.
data do: [:each |
    (each = aKey) ifTrue: [sum ←sum +1]].
(anInt < maxLevel) ifTrue: [
    upperTally ←(self upperChild: anId level: aLevel) localTally: aKey level: 1 root: anId,
    lowerTally ←(self lowerChild: anId level: aLevel) localTally: aKey level: 1 root: anId.
    ↑upperTally + lowerTally + sum].
↑sum.

Figure 2.2: A Concurrent Tally Method

2.3 Concurrency

CST does not exclude the use of concurrency in performing a single method. A more sophisticated tally: method is shown in Figure 2.2. Here I use messages upperChild and lowerChild to embed a tree on the COs[3]. When a CO receives a tally: message it sends two localTally messages down the tree simultaneously. When the localTally messages reach the leaves of the tree, the replies are propagated back up the tree concurrently. The new TallyCollection can still process many messages concurrently, but now it uses concurrency in the processing of a single message as well.

The use of a comma, ',', rather than a period, '.', at the end of a statement indicates that the method need not wait for a reply from the send implied by that statement before continuing to the next statement. When a statement is terminated with a period, '.', the method waits for all pending sends to reply before continuing.

---

[3]The implementation of methods upperChild and lowerChild is straightforward and will not be shown here.

class	Interval	*the class name*
superclass	Object	*the name of its superclass*
instance variables	l	*lower bound*
	u	*upper bound*
class variables		*none*
locks	rwLock	*implements readers and writers*

class methods

   l: aNum u: anotherNum   *creates a new interval*
     | newInterval |
     newInterval ←self new.
     newInterval l: aNum,
     newInterval u: anotherNum.
     ↑newInterval

   other class methods ...

instance methods

   contains: aNum   *tests for number in interval*
     require rwLock.
     | lin uin |
     lin ←l $\leq$ aNum,
     uin ←u $\geq$ aNum.
     ↑(lin and: uin)

   other instance methods ...

Figure 2.3: Description of Class Interval

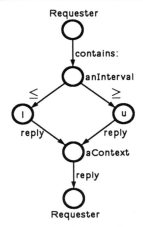

Figure 2.4: Synchronization of Methods

A simpler example of concurrency is shown in Figure 2.3. This figure shows a portion of the definition of Class Interval[4]. The definition has two methods; l:u: is a class method that creates a new interval, and contains: is an instance method that checks if a number is contained in an interval.

As shown in Figure 2.4, the contains: method is initiated by sending a message, contains: aNum, to object, anInterval, of class Interval. Objects of class Interval have two *acquaintances*[5] , l and u. To check if it contains aNum, object anInterval sends messages to both l and u asking l if $l \leq$ aNum, and asking u if $u \geq$ aNum. After receiving both replies, anInterval replies with their logical and.

Observe that the contains: method requires three actions. The first action occurs when the contains: message is received by anInterval. This action sends messages to l and u and creates a context, aContext, to which l and u will reply. The first reply to aContext triggers the second action which simply records its occurrence and the value in the reply. The second reply to aContext triggers the final action which computes the result and replies to the original sender. In this example the context object is used to join two concurrent streams of execution.

---

[4]The term Interval here means a closed interval over the real numbers, $\{a \in \Re \mid l \leq a \leq u\}$. This differs from the Smalltalk-80 [53] definition of class Interval.

[5]In the parlance of actor languages [1] an object, A's, acquaintances are those objects to which A can send messages.

Only the first action of the contains: method is performed by object anInterval. The subsequent actions are performed by object aContext. Thus, once the first action is complete anInterval is free to accept additional messages. The ability to process several requests concurrently can result in a great deal of concurrency. This simple approach to concurrency can cause problems, however, if precautions are not taken to exclude incompatible methods from running concurrently.

2.4 Locks

Some problems require that an object be capable of sending messages and receiving their replies while deferring any additional requests. In other cases we may want to process some requests concurrently, while deferring others. To defer some messages while accepting others requires the ability to *select* a subset of all incoming messages to be received. This capability is also important in database systems, where it is referred to as *concurrency control* [135].

Consider our example object, anInterval. To maintain consistency, anInterval must defer any messages that would modify I or u until after the contains: method is complete. On the other hand, we want to allow anInterval to process any number of contains: messages simultaneously.

SAL, an actor language, handles this problem by creating an *insensitive actor* which only accepts *become* messages [1][6]. The insensitive actor buffers new requests until the original method is complete. Lang's concurrent SIMULA [81] incorporates a *select* construct to allow objects to select the next message to receive. While exclusion can be implemented using select, Lang's language treats each object as a critical region, allowing only a single method to proceed at a time. Neither insensitive actors nor critical regions allow an object to selectively defer some methods while performing others concurrently.

Adding *locks* to objects provides a general mechanism for concurrency control. A lock is part of an object's state. Locks impose a partial order on methods that execute on the object. Each method specifies two possibly empty sets of locks: a set of locks the method *requires*, and a set of locks the method *excludes*. A method is not allowed to begin execution until all previous methods executing on the same object that exclude a required lock or require an excluded lock have completed. The concept of locks is similar to that of triggers [92].

---

[6]CST objects could use the Smalltalk become: message to implement insensitive actors.

A solution to the readers and writers problem is easily implemented with this locking mechanism. All readers exclude rwLock, while all writers both require and exclude rwLock. Many reader methods can access the object concurrently since they do not exclude each other. As soon as a writer message is received, it excludes new reader methods from starting while it waits for existing readers to complete. Only one writer at a time can gain access to the object since writers both require and exclude rwLock. This illustrates how mutual exclusion can also be implemented with a single lock.

2.5 Blocks

Blocks in CST differ from Smalltalk-80 blocks in two ways.

- A CST block may specify local variables and locks in addition to just arguments. [:arg1 :arg2 | (locks) :var1 :var2 | code]

- It is possible to break out of a CST block without returning from the context in which the value message was sent to the block. The down-arrow symbol, '↓', is used to break out of a block in the same way that '↑' is used to return out of a block.

Sending a block to a collection can result in concurrent execution of the block by members of the collection. Giving blocks local variables allows greater concurrency than is possible when all temporary values must be stored in the context of the creating method. Locks are provided to synchronize access to static variables during concurrent execution.

2.6 Performance Metrics

Performance of sequential algorithms is measured in terms of time complexity, the number of operations performed, and space complexity, the amount of storage required [2]. On a concurrent machine we are also concerned with the number of operations that can be performed concurrently.

The algorithms and data structures developed in this thesis are based on a message-passing model of concurrent computation. Message-passing concurrent computers are communication limited. The time required to pass messages dominates the processing time, which we will ignore.

In sharp contrast, most existing concurrent algorithms have been developed assuming an ideal shared-memory multiprocessor. In the shared-memory model, communication cost is ignored. Processes can access any memory location with unit cost, and an unlimited number of processes can access a single memory location simultaneously. Performance of algorithms analyzed using the shared-memory model does not accurately reflect their performance on message-passing concurrent computers.

Communication cost has two components:

latency: the delay of delivering a single message in isolation,

throughput: the amount of message traffic the communication network can handle per unit time.

For purposes of analysis I will ignore throughput and consider only latency.

The programs in this thesis are analyzed assuming a binary n-cube interconnection topology. Programs are charged one unit of time for each communication channel traversed in a binary n-cube.

2.7 Summary

In this chapter I have introduced Concurrent Smalltalk (CST), a programming notation for message-passing concurrent computers. Its novel features include locks for concurrency control, and the ability to create distributed objects. CST borrows its syntax, late-binding, and inheritance directly from the Smalltalk programming language [53]. Many of the ideas in CST are borrowed from Athas' language, XCPL [3].

Distributed objects are implemented as a collection of constituent objects (COs). Any CO can receive a message sent to the distributed object. Since many COs can receive messages at the same time, a distributed object can process many messages simultaneously. The constituents of a distributed object are assigned to processing nodes when the object is created. Thus, distributed objects provide a mechanism for mapping a data structure onto an interconnection topology. Distributed objects are the foundation upon which concurrent data structures, such as the balanced cube described in Chapter 3, are built.

CST permits methods to exploit concurrency by sending several messages before waiting for any replies. CST also allows some methods to terminate without sending any reply. Thus a message can be forwarded across many objects before a reply is finally sent to the original requester.

CST methods are compiled into sequences of primitive actions that can be described using a behavior function (2.1). Context objects are used to hold the state of a method between actions and to join concurrent streams of execution. Primitive object behaviors can be implemented using the message-passing process model of computation [125]. However, as we will see in Chapter 5, a direct hardware implementation of the behavior function results in improved performance.

Chapter 3

The Balanced Cube

Sequential computers spend a large fraction of their time manipulating ordered sets of data. For these operations to be performed efficiently on a concurrent computer, a new data structure for ordered sets is required. Conventional ordered set data structures such as heaps, balanced trees, and B-trees [2] have a single root node through which all operations must pass. This root bottleneck limits the potential concurrency of tree structures, making them unable to take advantage of the power of concurrent computers. Their maximum throughput is $O(1)$. This chapter presents a new data structure for implementing ordered sets, the balanced cube [21], which offers significantly improved concurrency.

The balanced cube eliminates the root bottleneck allowing it to achieve a throughput of $O(\frac{N}{\log N})$ operations per unit time[1]. Concurrency in the balanced cube is achieved through uniformity. With the exception of the balancing algorithm, all nodes are equals. An operation may originate at any node and need not pass through a root bottleneck as in a tree structure. In keeping with the spirit of a homogeneous machine, the balanced cube is a homogeneous data structure.

Why is a concurrent data structure such as the balanced cube needed? Many applications are organized around an ordered set data structure. By using a balanced cube to implement this data structure, the application can be made concurrent with very little effort. The application is divided into partitions that communicate by storing data in and retrieving data from the balanced cube. Because the balanced cube can process these requests concurrently, accesses to the balanced cube do not serialize the application. In Section 3.8 we will see how a balanced cube can be used in a concurrent computer mail system,

---

[1] Unless otherwise specified, all logarithms are base two.

in a concurrent artwork analysis program, and in a concurrent directed-search algorithm.

The balanced cube's topology is well matched to binary n-cube multiprocessors. The balanced cube maps members of an ordered set to subcubes of a binary n-cube. A Gray code mapping is used to preserve the linear adjacency of the ordered set in the Hamming distance adjacency of the cube.

Previous work on concurrent data structures has concentrated on reducing the interference between concurrent processes accessing a common data base but has not addressed the limited concurrency of existing data structures. Kung and Lehman [78] have developed concurrent algorithms for manipulating binary search trees. Lehman and Yao [83] have extended these concepts and applied them to B-trees. Algorithms for concurrent search and insertion of data in AVL-trees [33] and 2-3 trees [34] have been developed by Ellis. Ellis has also developed concurrent formulations of linear hashing [35] and extendible hashing [36].

These papers introduce a number of useful concepts that minimize locking of records, postpone operations to be performed, and use marking mechanisms to modify the data structure. However, these papers consider the processes and the data to be stationary, and thus do not address the problems of moving processes and data between the nodes of a concurrent computer. The cost of communications, which we assume to dominate processing costs, has largely been ignored.

The remainder of this chapter describes the balanced cube and how it addresses the issues of correctness, concurrency, and throughput. In the next section the data structure is presented, and the consistency conditions are described. The VW search algorithm is described in Section 3.2. VW search uses the distance properties of the Gray code to search the balanced cube for a data record in $O(\log N)$ time while locking only a single node at a time. An insert algorithm is presented in Section 3.3. Insertion is performed by recursively splitting subcubes of the balanced cube. Section 3.4 discusses the delete operation. Deletion is accomplished by simply marking a record as deleted. A background garbage collection process reclaims deleted subcubes. The insertion and deletion algorithms tend to unbalance the cube. A balancing algorithm, presented in Section 3.5, acts to restore balance. Each of the algorithms presented in this chapter is analyzed in terms of complexity, concurrency, and correctness. Section 3.6 extends the balanced cube concept to B-cubes which store several records in each node. Section 3.7 discusses the results of experiments run to verify the balanced cube algorithms. The chapter concludes with a discussion of some possible balanced cube applications in Section 3.8.

3.1 Data Structure

3.1.1 The Ordered Set

An *ordered set* is a set, S, of objects on which a linear ordering $<$ has been defined, $\forall a, b \in S$ either $a < b$ or $b < a$ and $a = b$ unless a and b are the same object. In many applications these objects are records and the linear order is defined by the value of a key field in each record. In this context the ordered set is used to store a database of relations associating the key field with the other fields of the record. The order relation defined on the keys of the records is implicit in the structure. A data structure implementing the ordered set must efficiently support the following operations.

at: key   return the object associated with a key.

at: key put: object   add an object to the set

delete: key   remove the object associated with key from the set.

from: lkey to: ukey do: aBlock   concurrently send a value: message to aBlock for each element of the set of objects with keys in the range [lkey,ukey].

succ: key1   return the object with the smallest key greater than key1.

pred: key1   return the object with the largest key smaller than key1.

max   return the maximum object.

min   return the minimum object.

In this chapter we will restrict our attention to developing algorithms for the search (at:), insert (at:put:) and delete operations. The remaining functions can be implemented as simple extensions of these three fundamental operations. The succ: and pred: operations can be implemented using the nearest neighbor links present in the balanced cube.

3.1.2 The Binary n-Cube

The balanced cube is a data structure for representing ordered sets that stores data in subcubes of a binary n-cube [98], [126]. A binary n-cube has $N = 2^n$ nodes accessed by n-bit addresses. Each bit of the address corresponds to a

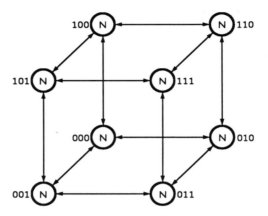

Figure 3.1: Binary 3-Cube

dimension of the cube. The node or subcube with address a_i is denoted $N[a_i]$. If the address is implicit, the node will be referred to as N. The binary n-cube is connected so that node $N[a_i]$ is adjacent to all nodes whose addresses differ from a_i in exactly one bit position: $\{a_i \oplus 2^j \mid 0 \leq j \leq n-1\}$. A binary 3-cube with nodes labeled by address is shown in Figure 3.1.

An m-subcube of a binary n-cube is a set of $M = 2^m$ nodes whose addresses are identical in all but m positions. An m-subcube is identified by an address that contains unknowns, represented by the character X, in the m bit positions in which its members' addresses may differ. For example, in Figure 3.1 the top of the 3-cube is the 1XX subcube. The top front edge is the 1X1 subcube.

A right m-subcube is an m-subcube which has unknowns in the least significant m bits of the address. No X is to the left of a 0 or 1 in a right subcube address. For example, the 1XX subcube is a right subcube while the 1X1 subcube is not. A node is a right 0-subcube, a singleton set, since it has zero Xs in its address. The *corner node* of a right subcube $N[a]$ is the node with the lowest address in the subcube, $N[\min(a)]$. The corner node address is the subcube address with all unknown bits set to zero. The *upper nodes* of a right subcube $N[a]$ are all the nodes in the subcube other than the corner node: the elements of the set $N[a] \setminus N[\min(a)]$.

3.1.3 The Gray Code

The balanced cube uses a Gray code [58] to map the elements of an ordered set to the vertices of a binary n-cube. Consider an integer, I, encoded as a weighted binary vector, $b_{n-1}, \ldots, b_0$, so that

$$I = \sum_{j=0}^{n-1} b_j 2^j. \tag{3.1}$$

The reflected binary code or Gray code representation of I, a bit vector $G(I) = g_{n-1}, \ldots, g_0$, is generated by taking the modulo-2 sum of adjacent bits of the binary encoding for I [58].

$$g_i = \begin{cases} b_i \oplus b_{i+1} & \text{if } i < n-1 \\ b_i & \text{if } i = n-1 \end{cases} \tag{3.2}$$

Since the $\oplus$ operation is linear, we can convert back to binary by swapping g_i and b_i in equation 3.2. We use the function $B(J)$ to represent the binary number whose Gray code representation is J.

$$b_i = \begin{cases} g_i \oplus b_{i+1} & \text{if } i < n-1 \\ g_i & \text{if } i = n-1 \end{cases} \tag{3.3}$$

By repeated substitution of equation 3.3 into itself we can express b_i as a modulo-2 summation of the bits of $G(I)$.

$$b_i = \sum_{j=i}^{n-1} g_j \ (\text{mod } 2) \tag{3.4}$$

While these equations serve as a useful recipe for converting between binary and Gray codes, we gain more insight into the structure of the code by considering a recursive list definition of the Gray code. For any integer, n, we can construct a list of $N = 2^n$ integers, gray(n), so that the Ith element of gray(n) is an integer whose binary encoding is identical to the Gray encoding of I. The construction begins with the Gray code of length 1. At the ith step we double the length of the code by appending to the current list a reversed copy of itself with the ith bit set to one[2].

---

[2]In (3.5) [0] denotes the list containing the number zero. The function append(x,y) in (3.6) appends lists x and y. The function reverse(z) reverses the order of list z. Also in (3.6) the addition is performed with scalar extension. The number 2^{n-1} is added to every element of the reversed list.

$$\text{gray}(0) = [0]. \tag{3.5}$$

$$\text{gray}(n) = \text{append}(\,\text{gray}(n-1)\,,\, 2^{(n-1)} + \text{reverse}(\text{gray}(n-1))\,). \tag{3.6}$$

It is this reversal that gives the code the symmetry and reflection properties that we will use in developing the balanced cube search algorithm.

In the linear space of the ordered set, element I is adjacent to elements $I \pm 1$. In the cube space, however, the distance between two nodes is the Hamming distance between the node addresses: the number of bit positions in which the two addresses differ. For nodes A and B to be adjacent, they must be Hamming distance one apart, $d_H(A, B) = 1$. The Hamming distance between I and $I-1$, $d_{HA}(I)$ is given by the recursive equation.

$$d_{HA}(I) = \begin{cases} d_{HA}(\frac{I}{2}) + 1 & \text{if } 2|I,\ I=0 \\ 1 & \text{if } 2 \,|I \\ \text{undefined} & \text{if } I=0 \end{cases} \tag{3.7}$$

A plot of this function is shown in Figure 3.26 on page 67.

For example, in the case where $I = \frac{N}{2}$ and $I-1 = \frac{N}{2} - 1$, the elements are at opposite corners of the cube, distance n apart. The Gray code has the property that $d_H(G(I), G(I+1)) = 1$, $\forall I \ni 0 \leq I \leq (N-2)$. Thus, if we map element I of the linear order to node $G(I)$ of the binary n-cube, nodes that are adjacent in linear space are also adjacent in cube space. A Gray code mapping of integers onto a binary 3-cube is shown in Figure 3.2.

3.1.4   The Balanced Cube

In a balanced cube, each datum is associated with a right subcube, $N[a_i]$, and is stored in a constituent object in the corner node, $N[\min(a_i)]$, of the subcube. Figure 3.3 shows the header for class Balanced Cube. A datum is composed of a key, N key, an object associated with the key, N object, the dimension of the subcube, N dim, and a flag, N flag, that indicates the status of the subcube. The data are ordered so that if $B(a_1) > B(a_2)$, $N[a_1]$ key $\geq N[a_2]$ key. Node addresses are ordered using the inverse Gray code function; thus, if two addresses are adjacent in the order, they will also be Hamming distance one apart.

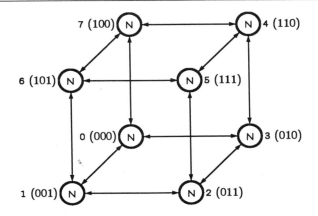

Figure 3.2: Gray Code Mapping on a Binary 3-Cube

class	Balanced Cube	*the class name*
superclass	Distributed Object	
instance variables	key	*defines the order*
	data	*object associated with key*
	dim	*the dimension of the subcube*
	flag	*status of subcube*
class variables		*none*
locks	rwLock	*implements readers and writers*

Figure 3.3: Header for Class Balanced Cube

For the remainder of this chapter I will refer to both cube addresses and to linear order addresses. A cube address is the physical address of a processing node. The parenthesized binary numbers in Figure 3.2 are cube addresses. A linear address is the position of a node in the linear order. For example, the integers (0-7) in Figure 3.2 are linear addresses. Linear addresses A_{lin} are related to cube addresses A_{cube} by (3.2) and (3.4).

$$A_{\text{lin}} \;=\; B(A_{\text{cube}})$$
$$A_{\text{cube}} \;=\; G(A_{\text{lin}})$$

$$(3.8)$$

Upper nodes of the subcube $N[a_i]$ are flagged as slaves to the corner node by setting N flag $\leftarrow \#$slave. Any messages transmitted to an upper node $N[a_u]$ are routed to the corner node of the subcube to which $N[a_u]$ belongs. There is one exception to this routing rule. A split message is always accepted by its destination and never forwarded. This message is the mechanism by which upper nodes become corner nodes. Since the cube is balanced, most corner nodes have dimensions differing only by a small constant. Thus, the message routing time between adjacent corner nodes will be limited by a small constant.

Data are associated with the subcubes rather than the nodes of a binary n-cube to allow ordered sets of varying sizes to be mapped to a cube of size 2^n. For example, a singleton set mapped to the 3-cube of Figure 3.1 would be associated with the subcube XXX, the entire cube. If a second element is added to the set, the cube will be split. One element will be associated with the $0XX$ subcube and the other element with the $1XX$ subcube. This splitting is repeated as more elements are added to the set.

A balanced cube is balanced in the sense that in the steady state, the dimensions of any two subcubes of the balanced cube will differ by no more than one. This degree of balance guarantees $O(\log N)$ access time to any datum stored in the cube. The balance condition is valid only in the steady state. Several insert or delete operations in quick succession may unbalance the cube. A balancing process which runs continuously acts to rebalance the cube.

There are two consistency conditions for a balanced cube. It must be ordered as described above and operations on the cube must be serializable giving results consistent with sequential execution of the same operations ordered by time of completion. This condition guarantees correct results from concurrent operations.

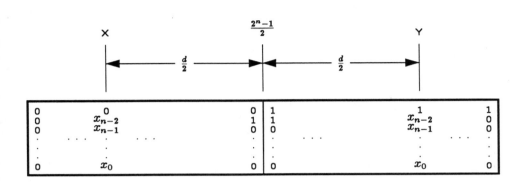

Figure 3.4: Calculating Distance by Reflection

3.2 Search

3.2.1 Distance Properties of the Gray Code

To develop a search algorithm for the balanced cube, we need to know the distance properties of the Gray code; that is, for any element of the ordered set mapped onto the cube, at what distance in linear space its neighbors are in cube space. The distance properties of the mapping tell us how much we can reduce the (linear) search space with each nearest neighbor query in the cube. To achieve $O(\log N)$ search time we must cut the search space in half with no more than a constant number of messages.

The reflection properties of the Gray code give us an easy method of calculating distance in a balanced cube. Consider some node, X, in a balanced n-cube. As shown in Figure 3.4, if we toggle the most significant bit of node address X, we generate address $Y = X \oplus 2^{n-1}$. In linear space, Y is the reflection of X through $\frac{2^n - 1}{2}$. Thus, the linear distance between node X and its neighbor, Y, in the $n - 1^{\text{st}}$ dimension is

$$d_{LN}(X, n-1) = 2 \left| X - \frac{2^n - 1}{2} \right|. \tag{3.9}$$

To calculate the distance in a lower dimension, say k, we reflect about the center of the local gray(k) list. Thus, the linear distance from a node with address X to its neighbor in the kth dimension is given by

$$d_{LN}(X, k) = 2 \left| (X(\text{mod } 2^{k+1})) - \frac{2^{k+1} - 1}{2} \right|. \tag{3.10}$$

The tables below show the distance function (3.10) for each dimension, k, of a balanced 4-cube. In each dimension, k, the first table shows the cube address, $G(X)$ for the Xth element in the linear order. The second table lists the neighbor of each node, X, in the kth dimension, $N(X, k)$. The third table shows the distance to this neighbor, $d_{LN}(X, k) = |N(X, k) - X|$. To find X's neighbor in dimension k, we convert X to a cube address, $G(X)$, toggle the kth bit, $G(X) \oplus 2^k$, and convert back to the linear order, $N(X, k) = B(G(X) \oplus 2^k)$. For example, the neighbor of node $X = 4$ in dimension $k = 1$ is node $N(4, 1) = 7$. The distance to this node is $d_{LN}(4, 1) = |7 - 4| = 3$.

X	0	1	2	3	4	5	6	7	8	9	10	11	12	13	14	15
$G(X)$	0	1	3	2	6	7	5	4	12	13	15	14	10	11	9	8

X	0	1	2	3	4	5	6	7	8	9	10	11	12	13	14	15
$N(X, 0)$	1	0	3	2	5	4	7	6	9	8	11	10	13	12	15	14
$N(X, 1)$	3	2	1	0	7	6	5	4	11	10	9	8	15	14	13	12
$N(X, 2)$	7	6	5	4	3	2	1	0	15	14	13	12	11	10	9	8
$N(X, 3)$	15	14	13	12	11	10	9	8	7	6	5	4	3	2	1	0

X	0	1	2	3	4	5	6	7	8	9	10	11	12	13	14	15
$d_{LN}(X, 0)$	1	1	1	1	1	1	1	1	1	1	1	1	1	1	1	1
$d_{LN}(X, 1)$	3	1	1	3	3	1	1	3	3	1	1	3	3	1	1	3
$d_{LN}(X, 2)$	7	5	3	1	1	3	5	7	7	5	3	1	1	3	5	7
$d_{LN}(X, 3)$	15	13	11	9	7	5	3	1	1	3	5	7	9	11	13	15

The distance functions shown in these tables are plotted in Figure 3.5. The symmetry of reflection is clearly visible. In each dimension, k, we have 2^{n-k-1} Vs centered on right subcubes of dimension $k + 1$. There are eight Vs of dimension 0, four Vs of dimension 1, two Vs of dimension 2, and one V of dimension 3. For example, the nodes at linear addresses 4-7 constitute a V of dimension 1. Combining this V with it neighboring V (addresses 8-11) gives addresses 4-11, a W of dimension 1.

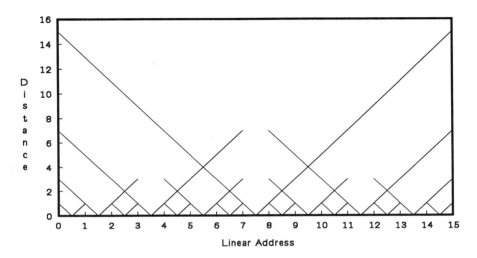

Figure 3.5: Neighbor Distance in a Gray 4-Cube

Definition 3.1 A *V* of dimension k is a right subcube of dimension $k + 1$: a collection of 2^{k+1} nodes beginning on a multiple of 2^{k+1} in the linear order.

Definition 3.2 A *W* of dimension k is two adjacent Vs of dimension k.

We use these Vs and Ws in the following section to develop a new search algorithm.

3.2.2 VW Search

VW search finds a search key in the Gray cube by traversing the Vs and Ws of the distance function shown in Figure 3.5. The neighbors of a node, X, are those nodes that are directly across a V from X in Figure 3.5. The search procedure sends messages across these valleys, selecting a search path that guarantees that the search space is halved every two messages.

Messages:

VW search is performed by passing messages between the nodes of the cube being searched. The body of the search uses two messages: **vSearch** and **wSearch**.

When a node receives one of these search messages, it updates the state fields of the message and forwards it to the next node in the search path. Nodes never wait for a reply from a message. The formats of the search messages are shown below. The search state is represented by the destination node, two dimensions: vDim and wDim, and a search mode: V or W.

> vSearch: aKey vDim: vDim wDim: wDim
> wSearch: aKey vDim: vDim wDim: wDim

In VW search we encode the search space into the destination address, self, and a dimension, wDim. wDim is the dimension of the smallest W in the distance function which contains the search space. A second dimension, vDim, is the dimension of the smallest V which completely contains our current W and thus the search space. vDim can be computed from wDim and self; however, it is more convenient to pass it in the message than to recompute it at each node.

The wDim, self encoding of the search space can be converted to the conventional upper bound, lower bound (U, L) representation by means of the reflect function. From (3.10) we know that the reflection in the linear space about dimension, d, of node X is given by

$$f_R(X, d) = X - 2(X(\bmod 2^{d+1})) + 2^{d+1} - 1. \tag{3.11}$$

The current position, self, or its reflection in the wDim dimension is one bound of the search space, and the reflection of this bound in the vDim dimension is the other bound. Thus, if the current address is S, the wDim is W, and the vDim is V, we can calculate the linear bounds of the search space (L, U) from

$$L(S, W, V) = \min(S, f_R(S, W), f_R(S, V), f_R(f_R(S, W), V)), \tag{3.12}$$

$$U(S, W, V) = \max(S, f_R(S, W), f_R(S, V), f_R(f_R(S, W), V)). \tag{3.13}$$

Algorithm:

VWsearch operates by passing vSearch and wSearch messages between the nodes of a balanced cube. Each message reduces the search space by comparing the search key to the key stored in the destination node.

When a node receives a vSearch message, the search space extends between the current node's neighbors in the V and W dimensions (N_V and N_W) as shown in

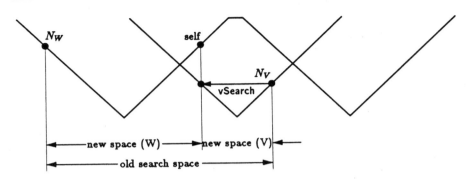

Figure 3.6: Search Space Reduction by vSearch Method

Figure 3.6. These neighbors will always be in opposite directions. By examining the key at the present node, the vSearch method makes the current node a new endpoint of the search space selecting N_V or N_W as the other endpoint. The dimension of the neighbor chosen becomes the new V dimension and the W dimension is decreased until a W neighbor in the appropriate direction is found.

When the W dimension has been reduced below the dimension of the current node, X, then X's W neighbor is contained within X's subcube. Thus there is no point in sending a message to the W neighbor, and the search is completed. Before terminating the search, however, X checks the contents of its linear neighbor in the direction of the key to verify that the key hasn't been inserted in the cube during the search. If the key isn't found, the search terminates with a nil reply. Otherwise, the search continues by increasing the W dimension above the dimension of X's subcube. The method for vSearch is shown in Figure 3.7 [3].

When a node receives a wSearch message, the search space extends from its W neighbor (N_W) to that neighbor's V neighbor (N_V) as shown in Figure 3.8. The wSearch method makes the current node one endpoint of the new search space, selecting between N_W and N_V as the other endpoint. If N_W is chosen as the endpoint, the search proceeds as in vSearch. If N_V is the endpoint, however, the dimension remains unchanged and a vSearch message is forwarded to the current node's V neighbor. The wSearch method is shown in Figure 3.9.

---

[3]The methods for neighbor:, upperNeighbor:, lowerNeighbor:, key:SameSideAsDim: and reduceDim:key: are omitted for the sake of brevity. Their implementation is straightforward.

instance methods for class Balanced Cube

    at: aKey   *reply the object associated with aKey*
    ||
    ↑self vSearch: aKey wDim: MaxWdim vDim: MaxVdim mode: vMode

    vSearch: aKey wDim: wDim vDim: vDim   *search for aKey*
        exclude rwLock.   *a reader method*
        |newVDim newWDim | *new dimensions of search space*
        (key = aKey) ifTrue:[requester reply: data].   *check if found*
        (self key: aKey sameSideAsDim: wDim) ifTrue: [
                newVDim ←wDim,
                newWDim ←wDim - 1.]
        ifFalse: [
                newVDim ←vDim,
                newWDim ←wDim.]
        newWDim ←self reduceDim: wDim key: aKey.
        (wDim < dim) ifTrue: [
                (key < aKey) ifTrue: [
                     (aKey < ((self upperNeighbor) key)) ifTrue: [requester reply: nil]]
                ifFalse: [
                     (aKey > ((self lowerNeighbor) key)) ifTrue: [requester reply: nil].
                newWdim ←self increaseDim: wDim key: aKey.]
        (self neighbor: newWDim) wSearch: aKey wDim: newWDim vDim: newVDim.

Figure 3.7: Methods for **at:** and **vSearch**

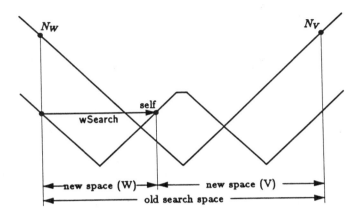

Figure 3.8: Search Space Reduction by wSearch Method

instance methods for class Balanced Cube

   wSearch: aKey wDim: wDim vDim: vDim   *search for aKey*
      exclude rwLock.
      | |
      (key = aKey) ifTrue:[requester reply: data].   *check if found*
      (self key: aKey sameSideAsDim: wDim)
         ifTrue: [self vSearch: aKey wDim: wDim vDim: vDim]
         ifFalse: [(self neighbor: vDim) vSearch: aKey wDim: wDim vDim: vDim.]

Figure 3.9: Method for wSearch

Example 3.1 The search technique is best described by means of an example. Consider the following table.

X	0	1	2	3	4	5	6	7	8	9	10	11	12	13	14	15
$G(X)$	0	1	3	2	6	7	5	4	12	13	15	14	10	11	9	8
Data	\$A	\$B	\$C	\$D	\$E	\$F	\$G	\$H	\$I	\$J	\$K	\$L	\$M	\$N	\$O	\$P

The table represents a Gray 4-cube where each node of the cube stores a single character symbol. Figure 3.10 shows the search of this Gray 4-cube for the key \$G stored at node $G(6)$. The search begins at node $G(2)$. The search is started with the message vSearch: \$G wDim: 5 vDim: 5. Since we know that the search key must be in the current dimension 5 trough of the W (this is the whole 4-cube), we start the search with a vSearch message. The subsequent search messages are as follows:

1. Since the search key, \$G, is greater than the key \$C stored at $G(2)$, node $G(2)$ sends the message wSearch: \$G wDim: 4 vDim: 5 to its dimension 4 neighbor, $G(13)$.

2. Since the key, \$G, is between $G(2)$'s key, \$C, and $G(13)$'s key, \$N, $G(13)$ sends the message wSearch: \$G wDim: 3 vDim: 4 to node $G(10)$.

3. The search key is not between \$N and \$K, so $G(10)$ must reflect the search (in the vDim dimension) to the other trough of the W by sending the message vSearch: \$G wDim: 3 vDim: 4 to node $G(5)$.

4. Since the key is not between $G(5)$'s key, \$F, and its neighbor $G(2)$'s key, \$C , the Wdim is decreased to find a neighbor in the direction of the key. $G(5)$ sends the message wSearch: \$G wDim: 2 vDim: 4 onto $G(6)$ where the search terminates successfully.

Example 3.2 Figure 3.11 shows two examples of searching a balanced cube which is not full and which is temporarily out of balance.

1. In the first example, a search for the contents of $G(4)$ is initiated from node $G(5)$ with the message vSearch: 4 wDim: 4 vDim: 4.

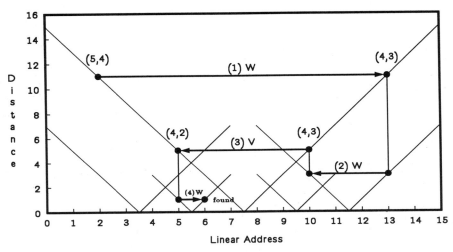

Figure 3.10: Example of VW Search

2. Since the search key is less than $G(5)$, node $G(5)$ forwards the search message to node $G(2)$, a slave node of node $G(0)$ with the message wSearch: 4 wDim: 3 vDim: 4. Since the search key is greater than the contents of $G(0)$ the W dimension is decremented to 0 and the message wSearch: 4 wDim: 0 vDim: 4 is sent to node $G(3)$. It is important to note that although $G(2)$ is a slave node and thus uses the value of the corner node, $G(0)$, the search continues from $G(2)$ and is not detoured to $G(0)$.

3. Node $G(3)$ is also a slave of $G(0)$ and thus less than the search key, so the search is reflected across the V dimension to node $G(4)$ with the message vSearch: 4 wDim: 0 vDim:4. The search key is found at node $G(4)$.

The second example in Figure 3.11 illustrates the case in which the search key is not present in the cube.

1-3. The search for the key, 3, is initiated at node $G(5)$. The search proceeds as above until the message vSearch: 4 wDim: 0 vDim: 4 reaches node $G(4)$.

4. To confirm that the key has not been inserted during the search, node $G(4)$ examines the key of its linear address neighbor, node $G(3)$ by sending $G(3)$ a key message.

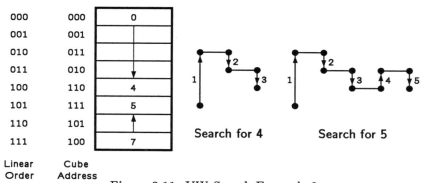

Linear Order	Cube Address	
000	000	0
001	001	
010	011	
011	010	
100	110	4
101	111	5
110	101	
111	100	7

Search for 4   Search for 5

Figure 3.11: VW Search Example 2

5. Node $G(3)$ replies with the value associated with its subcube, 0. Since 0 and the contents of $G(4)$, 4, bracket the search key, the search terminates by sending a nil reply to the original requester.

The remainder of this section analyzes the VW search algorithm to show that the order of the algorithm is $O(\log N)$ and to prove that the algorithm is deadlock free.

Lemma 3.1 Each execution of vSearch decreases wDim by at least 1.

Proof: There are two cases as shown in Figure 3.6:

1. If the search key is between the current node, self, and its W neighbor, wDim is explicitly decremented.

2. If the search key is between the current node, self, and its V neighbor, then the current W neighbor is in the wrong direction, so ReduceDim:key: will decrement wDim by at least 1 to find a neighbor in the proper direction.

Lemma 3.2 vSearch is executed at least once for every two search messages [4].

Proof: The only case in which vSearch is not executed is when a wSearch message is received and the key is not between the current node and its neighbor. The next message generated in this case is a vSearch message. Thus the vSearch method will be executed at least once for every two messages. ∎

Theorem 3.1 A VW Search of a Gray n-cube requires no more than $2(\log N + 1)$ messages.

Proof: From Lemmas 3.1 and 3.2 wDim is decremented at least once every two messages. Since wDim is initially $n = \log N$, after $2 \log N$ messages wDim will be zero. An additional two messages will either find the search key or decrement wDim below zero, causing termination. ∎

Theorem 3.2 The VW Search algorithm is deadlock free.

Proof: The VW Search algorithm locks only one node at a time: the one currently conducting the search. Since rwLock is never required, the key messages transmitted before terminating an unsuccessful search are never blocked. Thus, there is no possibility of deadlock. ∎

3.3   Insert

Messages:

The insert operation is initiated by sending an at:put: message to any node in the cube. This message starts a search of the cube for the insert key, aKey. When the search terminates, the data, anObject, is inserted by calling method localAt:put:. A split:key:data:flag: message is used by this method to split an existing right subcube into two right subcubes of lower dimension to make room for the insert.

```
at: aKey put: anObject
localAt: aKey put: anObject
split: aDim key: aKey data: anObject flag: aFlag
```

---

[4]Messages to self are local to the node and thus are not counted in this analysis.

instance methods for class Balanced Cube

localAt: aKey put: anObject   *insert after completing search*
```
    require rwLock exclude rwLock.
    | |
    (dim > 0) ifTrue: [
        dim ←dim - 1.
        (self key: aKey sameSideAsDim: dim) ifTrue: [
            (self neighbor: dim) split: dim key: aKey data: anObject flag: #valid]
        ifFalse: [
            (self neighbor: dim) split: dim key: key data: data flag: flag,
            key ←aKey,
            data ←anObject]
        requester reply: anObject]
    ifFalse: [
        requester reply: nil]
```

Figure 3.12: Method for localAt:put:

Algorithm:

The insert algorithm is identical to the search algorithm except that on completion, in addition to sending a reply, the insert splits a node and inserts the key and associated data. Rather than repeat the search algorithm here, only the changes will be described.

If the key being inserted is already in the cube, the insert replaces the object bound to the key with the object in the at:put: message. If the key being inserted is not already in the cube, the insert procedure must insert it. To do this, the not found reply of the search procedure listed above:

    requester reply: nil.

is replaced by a call to the method localAt:put: shown in Figure 3.12.

If the present node has a dimension greater than zero, then it is split by sending a split message to its upper half and decrementing its dimension. If the dimension is already zero, the insert terminates with a reply of nil. This does

---

instance methods for class Balanced Cube

   split: aDim key: aKey data: anObject flag: aFlag *splits a slave node from its parent*
      require rwLock exclude rwLock.

      | |
      key ←aKey,
      data ←anObject,
      dim ←aDim,
      flag ←aFlag.

Figure 3.13: Method for split:key:data:flag:

---

not necessarily mean that the cube is full. The cube may just be temporarily out of balance.

If the insert key and the linear order of the neighbor's address have the same relation to the current key and current address, the split message inserts the key and record into the corner node of the upper half subcube and sets its dimension to prevent it from routing further messages to the original corner node. The method belowNeighbor: dim returns true if the linear order address of the current node is less than the linear address of its neighbor in dimension dim. Figure 3.13 shows the split method. Once the dimension of the split node is set, the split is complete in that the split node will begin responding to messages rather than forwarding them to its corner node.

If the insert key and the linear order of the neighbor's address have opposite relations to the current key and current address, the split message copies the original corner node's key and record into the upper half subcube. The lower half subcube is then set with the new key and record. Note that between the assignment of the key and the assignment of the record to the lower half subcube, this subcube is in an inconsistent state.

To prevent an inconsistent state from being observed, both localAt: put: and split: key: data: flag: are *writer* methods. They both require and exclude rwLock. Thus, no other operation can be performed on the current node during an inconsistent state. This locking cannot cause deadlock, since the split node is in fact part of the locked node until the split is completed.

Consider splitting the subcube 000XXX into 0000XX and 0001XX. In the instant of time before the split, all nodes in 000XXX must route their messages to 000000. Immediately after the split, all messages to the upper half subcube 0001XX must be routed to 000100. For the cube algorithms to operate correctly, the split must be an atomic operation. Since the split occurs when the dimension of node 000100 is written, it is an indivisible operation. Before the dimension is written, messages to nodes 0001XX are routed to 000100 which forwards them to 00000 since it is not a corner node. After the dimension is written, these messages are accepted directly by 000100. Because the key and record of the split node are in fact not accessible before the dimension is updated, the split procedure does not have to require rwLock. This lock, however, makes the analysis of the operation simpler.

To prevent the possibility of simultaneously inserting the same key in the cube twice, it is necessary that the search terminate in the up direction unless the insert key is lower than the lowest key in the cube.

Example 3.3 Figure 3.14 shows the steps required to insert the key 3 into the cube of Figure 3.11. The search part of the insert proceeds as in Example 3.2 However, instead of terminating with a not found message, the key, 3, is inserted as follows:

1. Since the search must terminate in the UP direction, node $G(4)$ sends the search back to node $G(3)$. The state of the cube at this point is shown in Figure 3.14A.

2. As shown in Figure 3.14B, $G(0)$, the corner node of the $0XX$ subcube to which $G(3)$ belongs, decrements its dimension (from two to one), effectively detaching the $01X$ subcube, and sends a split message to its neighbor in dimension 1, node $G(3)$. $G(3)$ becomes the corner node of the newly formed subcube.

3. The split message inserts the key, 3, into node $G(3)$ and sets its dimension to 1 as shown in Figure 3.14C.

4. Finally, both nodes are unlocked as shown in Figure 3.14D.

Theorem 3.3 An insert operation in a stationary cube containing N nodes requires $O(\log N)$ time.

Proof: The initial stages of the insert are identical to the search operation and thus require $O(\log N)$ time. The final stage of the insert is the split operation which takes constant time. ∎

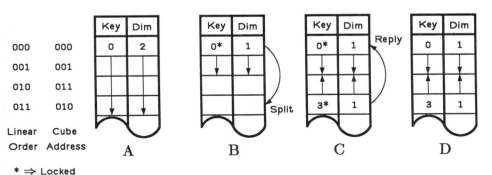

Figure 3.14: Insert Example

Theorem 3.4 An insert operation will not deadlock with other concurrent operations.

Proof: While the insert operation can lock out readers on two nodes simultaneously, the second node locked is part of the subcube which is locked by the first node. Placing the second lock operation does not increase the number of nodes which are locked. Rather, requiring and excluding rwLock in the split method assures that the upper half subcube will remain locked after its dimension is set to make it an independent subcube. This second subcube is in effect created by the insert and thus cannot previously have been locked by another operation. This node cannot be created by another operation during the final stage of the insert, since its corner node is locked, and the only way to create a node is to split it from its corner node. Thus, an insert operation will never have to wait to gain access to the split node. ∎

3.4 Delete

Messages:

The delete operation is initiated by sending a delete: message to any node in the cube. This message initiates a search for the node containing the delete key. If found, the operation marks this node as deleted and replies to the requester. After the node is marked deleted, it sends a mergeReq message to its *merge neighbor*. The merge neighbor merges with the deleted node to recover its space. The messages mergeUp, mergeDown, move, and copy are used to merge the two nodes.

The following is a list of the principal message selectors used to implement the delete: operation.

> **delete: aKey**
> **mergeReq: anId flag: aFlag dim: aDim**
> **mergeUp**
> **mergeDown: aKey data: anObject flag: aFlag**
> **move: aNode**
> **copy: aKey data: anObject flag: aFlag**

Algorithm:

The delete algorithm is identical to the search until the key is found. Then the node is marked deleted, flag ←#deleted, and a mergeReq message is sent to the deleted node's merge neighbor. This has the result of routing all messages addressed to this node except mergeUp, mergeDown, and copy messages to its merge neighbor.

Definition 3.3 The *merge neighbor* of a node, $N[a]$, with address, a, is the node $N[m(a)]$ with address, $m(a) = a \oplus 2^{N[a]\text{dim}}$. If the subcubes cornered by nodes $N[a]$ and $N[m(a)]$ are of the same dimension, they can be merged to form a subcube of greater dimension. Further, node $N[m(a)]$ is the only node with which node $N[a]$ can be merged.

When a node, A, receives a mergeReq message from another node, B, A determines if B is its merge neighbor by comparing dimensions. There are two possible cases, as shown in Figure 3.15. If the two nodes are of the same dimension (Figure 3.15A,B), they are merged. The merge is accomplished by node A's sending a mergeUp or mergeDown message to node B. If A is below B (Figure 3.15A) a mergeUp message is sent. A mergeDown message is sent if A is above B (Figure 3.15B). These messages have the effect of extending the subcube cornered by node A to include the subcube cornered by node B. The method invoked by a mergeReq message is shown in Figure 3.16.

When the two adjacent nodes A and B have different dimensions, a simple merge is not possible. This situation is shown in Figure 3.15C,D. Since A is the merge neighbor of B, it will always be the case that A dim $< B$ dim. In this case we copy the contents of node C, the linear address neighbor of node B, to node

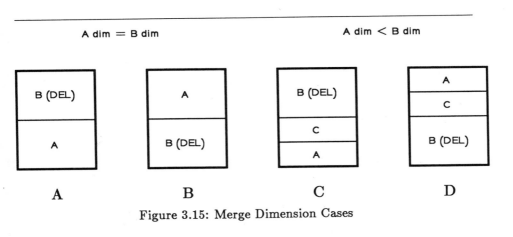

Figure 3.15: Merge Dimension Cases

B and mark C deleted. In performing the copy we reduce the dimension of the deleted subcube and make it possible for the linear address neighbor node, C, to merge subsequently with its merge neighbor A. The move: and copy:data:flag: messages are used to move the contents of node C to node B.

The merge operation combines the subcube cornered by node, A, with its adjacent subcube cornered by B. If the current node is the corner of the upper half subcube, the state of the current node is copied into the available lower half subcube with the mergeDown message. The method for mergeDown is shown in Figure 3.17. If this method is successful, the current node flag is set to #slave to indicate that it is no longer a corner node. Since the nodes are inconsistent while the copying takes place, this operation requires rwLock.

If the current subcube is below its adjacent subcube, then the current node is the corner of the combined subcube. In this case a mergeUp message is sent to the adjacent subcube to set its flag to #slave. Once this message completes successfully, the dimension of the current subcube is incremented to extend its domain over the merged subcube. The method for mergeUp is also shown in Figure 3.17.

Since a merge operation must lock both nodes A and B, a priority mechanism is used to prevent deadlock. If a mergeDown message arrives at a node which is locked, it terminates unsuccessfully. A mergeUp message will wait until the node is unlocked. The alternative 'or ↑false' in the lock specification for mergeDown causes it to return false rather than wait on an incompatible lock.

instance methods for class Balanced Cube

    mergeReq: anId flag: aFlag dim: aDim   *invoked after node anId is deleted*
       require rwLock exclude rwLock.
       | |
       (aFlag = #deleted) ifTrue: [
          (aDim = dim) ifTrue: [*same dimension, just merge*
            (anId > myId) ifTrue:[
               ((self co: anId) mergeUp) ifTrue:[dim ←dim + 1]]
            ifFalse[
               ((self co: anId) mergeDown: key data: data flag: flag) ifTrue:[flag ←#slave]]]
          ifFalse: [*smaller than neighbor, send move*
           (self neighbor: (aDim-1)) move: anId]]

Figure 3.16: Method for mergeReq:flag:dim:

Only a node's merge neighbor can send it a merge message; thus, there is only one case in which merge messages can form a cycle for resources. If two adjacent nodes of the same dimension, such as A and B in Figure 3.15A, are both deleted, these nodes will send mergeReq messages to each other. The mergeReq method will lock each node and send a mergeUp or mergeDown message to the other node. If the merge messages were both to wait on the locks, deadlock would occur. Instead, the mergeDown message terminates immediately. Its reply unlocks node B and allows the mergeUp message to proceed.

The messages move: and copy:data:flag: are used to move the contents of one node to another. When the move: message is received by a node, that node attempts to copy itself to the destination of the move by sending a copy: message to the destination. If the copy: succeeds, it replies to the move: which then marks the source node deleted. The methods for move and copy are shown in Figure 3.18.

Example 3.4 This example illustrates the simplest case of garbage collection, where the nodes are the same size and all that is required is a merge. Figure 3.19A shows the state of a 2-cube where the key stored in $G(3)$ has just been deleted. The following messages merge the deleted node with its neighbor.

---

instance methods for class Balanced Cube

mergeDown: aKey data: anObject flag: aFlag   *copy a node's state and absorb it*
require rwLock exclude rwLock or ↑false.
| |
key ←aKey, data ←anObject, flag ←aFlag.
↑true.

mergeUp   *merge with node below by becoming a slave*
exclude rwLock.   *a reader operation*
| |
flag ←#slave.
↑true.

Figure 3.17: Methods for mergeUp and mergeDown:data:flag:

---

instance methods for class Balanced Cube

move: anId   *attempt to move contents to node anId*
require rwLock exclude rwLock.   *a writer operation*
| |
((self co: anId) copy: key data: data flag: flag) ifTrue:[flag ←#deleted]

copy: aKey data: anObject flag: aFlag   *replace contents if deleted or free*
require rwLock exclude rwLock or ↑false.   *doesn't wait*
| |
((flag = #deleted) or: (flag = #free)) ifTrue: [
    key ←aKey, data ←anObject, flag ←aFlag.
    ↑true]
ifFalse:[↑false]

Figure 3.18: Methods for move: and copy:data:flag:

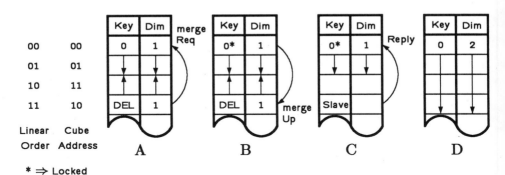

Linear Order	Cube Address
00	00
01	01
10	11
11	10

* ⇒ Locked

Figure 3.19: Merge Example: A dim $= B$ dim

1. To initiate collection, $G(3)$ sends a mergeReq message to its merge neighbor $G(0)$.

2. The mergeReq method locks $G(0)$ and sends a mergeUp message to $G(3)$ as shown in Figure 3.19B. This message locks $G(3)$. It will always succeed since mergeUp messages have priority over mergeDown messages.

3. As shown in Figure 3.19C, the mergeUp method sets $G(3)$'s flag equal to #slave effectively attaching it to the $0X$ subcube.

4. After the merge method replies, $G(0)$ increments its dimension to 2 to reflect the fact that the two subcubes, $1X$ and $0X$, have been merged to form a single subcube, XX. The final state of the subcube is shown in Figure 3.19D.

Example 3.5 Figure 3.20 illustrates the case where A dim $< B$ dim.

1. Node $G(3)/0$ (node $G(3)$ with dimension 0) receives a mergeReq message from node $G(0)/1$, as shown in Figure 3.20A.

2. The linear address neighbor of $G(0)/1$ is the neighbor of $G(3)$ in the dim - 1 dimension, $G(2)$. Node $G(3)$ sends a move:$G(0)$ message to $G(2)$ as shown in Figure 3.20B.

3. The move locks node $G(2)$ and copies the key, record and flag from node $G(2)$ to node $G(0)$ by sending a copy message as shown in Figure 3.20C.

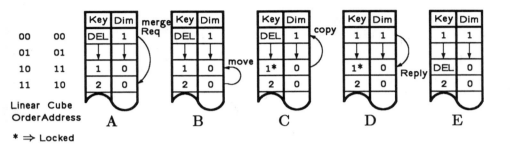

Figure 3.20: Merge Example: A dim $< B$ dim

4. When **copy** replies successfully to node $G(2)$ (Figure 3.20D) node $G(2)$ is marked deleted.

5. As illustrated in Figure 3.20E. Node $G(2)$ will now send a **mergeReq** to node $G(3)$ initiating equal dimension garbage collection.

Theorem 3.5 To delete a key from a cube with N nodes requires $O(\log N)$ time.

Proof: The search portion of the delete requires $O(\log N)$ time. Marking the node deleted and merging the node with its neighbor requires constant time. ∎

Theorem 3.6 The delete operation will not deadlock with other concurrent operations.

Proof: The delete operation locks only one node at a time. ∎

Theorem 3.7 The merge operations will not deadlock with other concurrent operations.

Proof: Although the merge operations lock two nodes simultaneously, this locking is ordered so that a node, A, will only wait for a node with an address greater than A to become unlocked. Thus, it is impossible to have a cycle of nodes waiting on each other's locks. ∎

Before proving that concurrent search, insert, delete, and merge operations will give the same result as running the operations sequentially in order of completion, we need to define some terms and prove one lemma about concurrency.

Definition 3.4 An operation *commits* when it has made a final decision to modify the state of a node in the cube and/or to reply with a particular result. Once an operation *commits* to modifying the state of a node, it must follow through and perform the modification. It cannot back out after committing.

Definition 3.5 The *commit condition* is the condition which must occur for an operation to commit.

Definition 3.6 An operation *completes* when it has finished modifying the state of a node. After an operation completes it cannot modify any additional state.

Definition 3.7 The *vulnerable period* of an operation is the period between the time it commits and the time it completes.

Definition 3.8 A *snapshot* of the cube is the state of all corner nodes of the cube with no methods in progress. Since there is no concept of simultaneity between nodes of the cube, each node may be stopped at any point as long as causality and order of completion are preserved.

Definition 3.9 The *neighborhood* of an operation includes all nodes whose states are examined by the operation between the time it commits and the time it completes.

Here are some examples:

- A search operation commits and completes at the same time. A successful search commits to replying with the data when it finds the requested key in the current node. An unsuccessful search commits to replying nil when it receives a reply from a linear address neighbor confirming that the search key is not in the cube.

- An insert operation commits when the search portion of the insert receives the reply from the query message to an adjacent node. The commit condition is that the present node and the adjacent node bracket the insert key. The insert operation completes when the split method unlocks its node. The node which is split constitutes the neighborhood of the insert operation.

- The commit condition for a delete operation is the key stored in the present node matching the delete key. When this condition is discovered, the operation commits. A delete is completed when the delete flag of the node is set true.

- A merge commits when the mergeUp or mergeDown message is accepted. The commit condition is that the two nodes being merged are adjacent. Completion occurs when the merged node is unlocked.

Lemma 3.3 If an operation, P's, commit condition is valid throughout P's vulnerable period and if P's neighborhood is not changed by another operation during this period, then any concurrent execution of P is consistent with a sequential execution of P ordered as follows:

- P is ordered after all operations R which complete before P commits.

- P is ordered before all operations S which commit after P completes.

- P is ordered either before or after any operation Q that completes during P's vulnerable period.

Proof: P's commit condition and P's neighborhood constitute the state of the cube which is visible to P. If this state remains constant from the time P commits to the time P completes, then P will act as if there were no concurrent operations, Q, during this period since it cannot see any changes caused by Q. It follows that P can be serialized with operations Q in any order. Since P's commit decision is valid after all operations R have completed, it will be valid if P is not started until after these operations have completed. Applying the same logic with S in place of P shows that operations S can be started after P completes without changing S's commit condition. ∎

Theorem 3.8 Concurrent search, insert, delete, and merge operations will give the same result as running the operations sequentially in order of completion.

Proof: The search, insert, delete and merge operations all meet the conditions in the hypothesis of Lemma 3.3:

Search completes at the time it commits and thus meets this condition.

The commit condition for insert is that the present node, A, and the node directly above the present node, B, straddle the key to be inserted, K. This condition always holds at completion since: (1) a new node C<K cannot be inserted between A and B since this insert would have to be performed at A and A is locked, and (2) if B is deleted during this period, then for any node D>B, D>K.

The commit decision for delete is that the delete key is found. The node containing this key is locked, so the condition still holds at completion.

The commit condition for merge is that the adjacent node is marked deleted and the merge operation is able to lock the node. Since both of the nodes being merged are locked during the vulnerable period, this condition is still valid when the operation completes. For all these operations, the neighborhood is the present node which is locked and thus remains constant during the critical period. ∎

3.5   Balance

The balancing process proceeds in three steps.

1. An imbalance between two adjacent subcubes, A and B, in the cube is recognized.

2. The subcube containing fewer data, say A, frees space on its border with B. Without loss of generality assume A is below B. To free space, the node containing the highest datum in A, AH, splits itself, freeing half its space.

3. The heavier subcube (containing more data), in this case B, moves its smallest datum to the space freed in step 2.

Imbalance is recognized by embedding a tree in the cube. As shown in Figure 3.21, for n=4, the tree is constructed by recursively dividing the cube into two subcubes. The node of each subcube closest in linear order to the other subcube is chosen as the corner node. This tree has one idiosyncrasy: messages

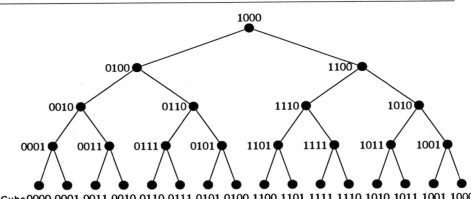

Figure 3.21: Balancing Tree, n = 4

to the outer child of a node[5] must traverse two communication links, while messages to the inner child of a node need to traverse only one link. Despite this shortcoming, however, the tree is ideal for balancing for two reasons. First, it evenly distributes the task of recognizing imbalance over all nodes of the cube except the zero node. Also, the root node of every cube is on the boundary of the cube across which a datum must be moved to balance the cube with an adjacent cube at the same level. Each root node participates in correcting an imbalance recognized by its parent.

The cube is balanced if, for each internal node in the tree, the number of keys stored in the subcubes represented by the two children of the node differ by less than 2 : 1. Using the number of keys in a subcube as the balancing criterion rather than the maximum or minimum dimension of a node in the subcube has the advantage that local imbalances are averaged out when considering global balance. This balance criterion also allows us to balance B-Cubes, described in Section 3.6.

---

[5]Here the *outer child* of a node, A, is the child of A to the outside of the subtree rooted by A's parent as drawn in Figure 3.21. The outer child of the root is the left child as drawn in Figure 3.21.

Messages:

Leaf and internal nodes periodically transmit size messages to their parent nodes. When the parent node receives the size message, it updates its size and checks the sizes of its two subcubes for imbalance.

If the root node of a subcube detects imbalance between the two halves of its subcube, it initiates balancing by moving records between its two children. This data transfer takes place in two steps. First, a free message is transmitted to the boundary node of the subcube containing fewer elements. This message causes the boundary node to split itself as in the insert operation, with the old key and record remaining in the node farthest from the subcube boundary. The boundary node of the subcube with the larger size is then sent a move message. This message locks the boundary node, copies its key and record to the freed node, and then marks the adjacent node deleted. The net effect is to move one datum from the larger subcube to the smaller subcube. While a node is marked free, it routes all its messages to the destination node. The root subcube repeats this operation until balance is restored to a 2:1 size ratio. It is important to note that because of the Gray code mapping, most of these messages traverse only a single link in the cube. The message from the root to its outer child is the only message that must traverse two links.

        size: anlnt of: anld
        free: anld

Algorithm:

The size method, shown in Figure 3.22, updates the size of self, checks for balance between its two subcubes, and possibly initiates balancing by sending a free message to the smaller of the two subcubes. The free method splits its destination subcube in half and sends a move message to the node in the other half subcube, instructing it to copy itself to the freed node and then to delete itself.

The free method, shown in Figure 3.23, is similar to insert in that it must split the present node to generate a free block. There are two cases. If the subcube contains more than one element, the boundary node is a corner node. Since it is right on the boundary, it must copy its present state into the split subcube and then free itself. If the subcube contains only a single element, the boundary node is a slave to the root which recognized the imbalance. In this case the root simply sends a split message to free the boundary half of its subcube. As with insert, locking two nodes simultaneously is permissible during a split

instance methods for class Balanced Cube

    size: anInt of: anId   *update size of subcube rooted at receiver*

```
    | |
    (myId < anId) ifTrue:[
        lowerSize ←anInt]
    ifFalse:[
        upperSize ←anInt].
    mySize ←lowerSize + upperSize. (lowerSize > (2 * upperSize)) ifTrue:[
        (self upperChild) free lowerChild]
    (upperSize > (2 * lowerSize)) ifTrue:[
        (self lowerChild) free upperChild]
```

Figure 3.22: Method for size:of:

since the two nodes were the same node at the time of the first lock, and it is impossible for another process to attempt to lock the split subcube after the original subcube is locked.

After a node is freed, the node which is to move to the freed subcube receives the move message. The move copies the boundary node's key and record to the freed node while preserving the freed node's dimension. After the copy completes, the boundary node is marked deleted. Although two nodes are locked simultaneously, unlike the merge operation, no priority resolution is required to prevent deadlock. Once a node is freed, there is only one node which can send a copy message to that node. Thus, as in the insert and free operations, for purposes of locking, the freed node is part of the boundary node from the moment it unlocks after being tagged free.

Example 3.6 Figure 3.24 shows a balancing operation on a 3-cube.

1. In Figure 3.24A, root node 100, $G(7)$, sees one record in the upper half of the cube and four records in the lower half of the cube. Recognizing this imbalance, $G(7)$ sends a free message to $G(4)$.

2. As shown in Figure 3.24B, since $G(4)$ is a slave to $G(7)$, the free operation locks $G(7)$, decrements its dimension, and sends a split message to $G(4)$.

instance methods for class Balanced Cube

    free: anId   *split self and send a move message to anId*

       require rwLock exclude rwLock

       | |

       (dim $>$ 0) ifTrue: [

          ((flag = #deleted) or: (flag = #free)) ifTrue:[(self co: anId) move: myId]]

       ifFalse:[

          dim ←dim - 1.

          (self adjacentTo: anId) ifTrue [

             (self neighbor: dim) split: dim key: key data: data flag: flag,

             flag ←free].

          (self co: anId) move: myId]

       ifFalse [

          (self neighbor: dim) split: dim key: key data: data flag: #free,

          (self co: anId) move: (myId xor: 2^{dim})]]

Figure 3.23: Method for free:

3. After the split message has marked $G(4)$ free, a move message is sent to $G(3)$ as shown in Figure 3.24C.

4. After the move completes, $G(3)$ is marked deleted and the cube is balanced as shown in Figure 3.24D.

The balancing operations alter none of the arguments in the proofs of Theorems 3.1 to 3.8 above. Thus, all of these theorems hold in a cube which is being dynamically balanced.

3.6   Extension to B-Cubes

A straightforward extension of the balanced cube is the B-cube. The B-cube is to a balanced cube what a B-tree is to a balanced tree. In the B-cube, rather than storing one record in each node, up to k records may be stored in each node. B-cube operations attempt to keep the number of records in each node between $\frac{k}{2}$ and k by splitting nodes when the number of records exceeds k and

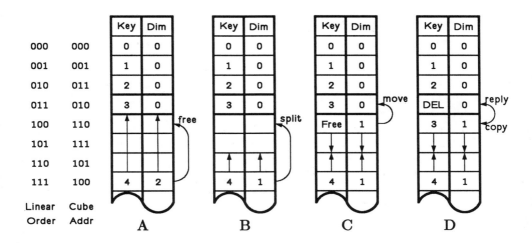

Figure 3.24: Balance Example

merging adjacent nodes when their combined number of records drops below $k + 1$. Within a B-cube node, records are sorted and searched by conventional means. Between nodes, the algorithms presented here for balanced cubes are applied with some modifications. For example, in the search procedure, a query message would reply with both upper and lower keys. The test for equality in this case would be *lower* $<=$ *key* $<=$ *upper*.

B-cubes have several advantages over balanced cubes:

- The overhead for maintaining the dimension and flag fields in each node is reduced. Rather than maintaining these fields for each record, their cost is spread out over up to k records. Locks in a B-cube can be either on a record basis or on a node basis. Write-locking at the node level and read-locking at the record level seem to make the most sense.

- In a B-cube, the majority of inserts and deletes can be performed entirely within a single node without splitting or merging. Thus, the number of node interactions is reduced. Also, balancing is required less frequently, since the number of operations which changes the node counts is reduced. Note, however, that when balancing is performed the amount of data to be moved has increased.

- It is expected that nodes will be swapped from a mass storage device. In the B-cube, the size of a node can be chosen to match a convenient transfer size for the storage device. In general, this size is larger than a single record.

A possible disadvantage of B-cubes is that they reduce the potential concurrency of the data structure. However, in most applications the number of records will greatly exceed the number of available processors, and the concurrency of B-cubes will not be the limiting factor. In fact, this reduction of concurrency is an advantage in the sense that it allows the granularity of the data structure to be smoothly varied over a large range.

3.7   Experimental Results

The balanced cube data structure has been implemented on a multiprocessor simulator, and a number of experiments have been performed to verify the correctness of the algorithms and to measure their throughput. The balanced cube simulator is a 3000-line C program [70]. The code is divided fairly evenly into three parts:

- A binary n-cube simulator which provides the message passing environment of a concurrent computer.

- The balanced cube algorithms.

- Instrumentation code to configure the cube simulator and to measure the performance of the balanced cube algorithms.

Two sets of experiments were run. The first set of experiments, described in detail in [20], was performed on an early version of the balanced cube which directly mapped the elements of the ordered set to the nodes of a binary n-cube. The current balanced cube algorithms, using a Gray code mapping, were used in the second set of experiments. After a few experiments were run to verify that the insert, delete, and balance operations consume only a modest portion of the cube's resources, all remaining experiments were performed using only the search operation.

Throughput experiments were run to determine if the data structure can achieve the predicted $O(\frac{N}{\log N})$ throughput. These experiments were run using a load model that applied a maximum uniform load to the cube. The experiments were run for both the direct mapped cube and the current balanced cube.

Throughput is the number of operations the data structure can perform per unit time. The balanced cube can perform N operations at a time and each operation requires $O(\log N)$ time, so the predicted throughput is $O(\frac{N}{\log N})$. In the steady state, the balanced cube can perform $O(\frac{N}{\log N})$ operations each message time.

The throughput results presented in this section assume a uniform load. Both the constituents to which requests are made and the keys searched for are uniformly distributed. A concentration of messages to one constituent or searching for a single key would cause a hot spot and reduce throughput. These throughput results also assume that data inserted into the balanced cube is uniformly distributed. If an adversary inserts a pathological sequence of data, balancing can, in the worst case, require $O(N)$ messages per operation reducing throughput to $O(1)$.

The throughput results for the original direct mapped cube of [20], shown in Figure 3.25, fail to achieve the predicted throughput. The direct mapped cube achieves a throughput of only $O(\frac{N}{\log^2 N})$.

The degradation of $O(\log N)$ is due to the non-uniformity of the Hamming distance between linear order neighbors, as expressed in Equation (3.7). The function, d_{HA}, can be thought of as a barrier function. Shown in Figure 3.26, this function represents how many channels a message between linear address neighbors must traverse. Degradation occurs because the channels corresponding to the higher barriers must carry more traffic than the channels corresponding lower barriers. Hence, these channels become congested.

The average barrier height is given by:

$$d_{HAL} = \frac{\sum_{i=1}^{n} i 2^{n-i}}{2^n} = \frac{2^{n+1} + 2}{2^n} \approx 2. \tag{3.14}$$

The degradation is the ratio of maximum barrier height to average barrier height or $\approx \frac{n}{2}$. The experimental data of Figure 3.25 agrees exactly with this figure.

The Gray code mapping used in the current balanced cube eliminates this degradation as shown in Figure 3.27. The throughput difference of $O(\log N)$ between Figures 3.25 and 3.27 illustrate the importance of developing data structures which match the topology of concurrent computers.

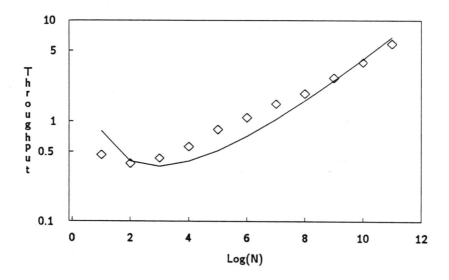

Figure 3.25: Throughput vs. Cube Size for Direct Mapped Cube. *Solid line is* $\frac{0.4N}{\log^2 N}$. *Diamonds represent experimental data.*

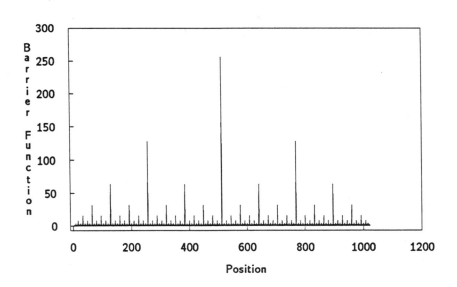

Figure 3.26: Barrier Function (n=10)

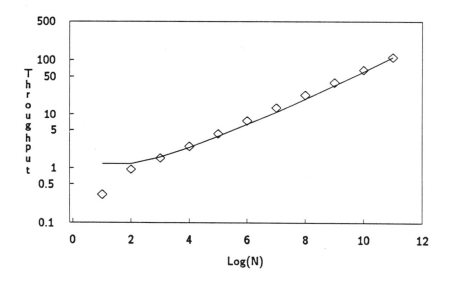

Figure 3.27: Throughput vs. Cube Size for Balanced Cube. *Solid line is* $\frac{0.6N}{\log N}$. *Diamonds represent experimental data.*

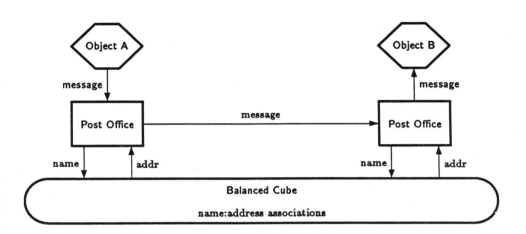

Figure 3.28: Mail System

3.8   Applications

A Mail System

Concurrent data structures such as the balanced cube provide a medium through which objects can communicate without knowing of each other's existence or physical location. Consider a mail system that forwards messages between objects that occasionally migrate from node to node. As shown in Figure 3.28, the mail system consists of a balanced cube used to hold the associations between object names and their current addresses, and local Post Offices that cache these associations and handle communications with objects. Objects interact through the Post Office rather than directly communicating with each other.

- When an object moves to a new node, it registers its new address by sending the message at: <name> put: <address> to its local Post Office. The Post Office inserts this association in the balanced cube.

- To send a message to an object, B, the sender object, A, transmits a message to its local Post Office. Each local Post Office maintains a cache of recently used object-address associations. If the address is not found in this cache, an at:<name> message is sent to the balanced cube to look up the address.

- If an address in the local cache is stale (the object has moved), the destination Post Office consults the balanced cube to find the correct address, forwards the message, and notifies the sending Post Office of the new address.

Using the PostOffice mechanism, objects can communicate without ever knowing anything about each other. Objects send messages to names. The object receiving messages for a given name can move or be replaced without notifying any of its customers. There is no central name server to become a bottleneck. The server that associates names with addresses is distributed and can process many requests simultaneously.

Artwork Analysis

Applications can be constructed by combining concurrent data structures. Consider the problem of integrated circuit *artwork analysis*. This problem has two aspects:

circuit extraction: discovering the electrical circuit of an integrated circuit from an examination of its layout geometry.

design rule checking: verifying that the layout obeys a set of geometrical design rules. These rules specify restrictions such as minimum feature width, minimum feature spacing, etc....

Traditionally, artwork analysis has been performed using a scan-line algorithm [5],[31],[16]. However scan line algorithms are inherently sequential as they involve traversing the chip in sequence from one end to the other. In this section we examine an approach to concurrent artwork analysis using balanced cubes.

The artwork for an integrated circuit is a set of polygons. Artwork analysis involves checking for interactions between polygons. An efficient algorithm must be selective in these checks to avoid the $O(N^2)$ complexity required to check every pair of polygons. If polygons are compared only with neighboring polygons, the number of comparisons can be significantly reduced.

To reduce the number of comparisons, we use a B-cube to maintain the spatial relationship between polygons in one dimension. Each polygon is enclosed in a bounding box, and the B-cube is ordered by the left x coordinate of the bounding box. Within each node of the B-cube, two indices into the local list of polygons are maintained, one ordered by x coordinate and one by y coordinate.

Artwork analysis is performed concurrently on this structure by having each polygon send a from: leftX to: rightX do: aBlock message to the B-cube. At each node of the B-cube, aBlock executes and, using the y index, selects only those polygons that overlap the sender in both coordinates. These polygons are then compared with the sender to check for design rule errors.

$O(N^{1.5})$ comparisons will be made on y coordinates, making this algorithm less efficient than $O(N \log N)$ sequential algorithms. This algorithm has the advantage, however, of being very concurrent, while the scan-line algorithms are inherently sequential.

By using a two-dimensional corner-stitched data structure as described in [95] it is possible to achieve concurrency without the $O(\sqrt{N})$ penalty imposed by ordering primarily in a single dimension. A corner-stitched data structure can be distributed by using the pointers as keys into a balanced cube. Since order is not required, the concurrent dictionary described in Appendix B could be used instead of the balanced cube.

Directed Search

Many problems involve the directed search of a state space. For example, most game-playing programs are built around an $\alpha - \beta$ search of a game tree that represents the state space of positions. The program begins from the current position and generates all possible successor positions. These successor positions are then expanded to generate positions two moves ahead and so on. At each step of the search there is a set of *active* positions: those positions that have been generated but not yet expanded. Active positions are expanded in order of their merit as determined by some *evaluation function*. Some positions may be pruned, eliminated from further consideration, on the basis of static evaluation functions.

We can construct a concurrent directed search algorithm by storing all generated positions in a balanced cube. As in the artwork analysis example above, a B-cube is used. Some hash function of position is used as a key to insert positions into the B-cube. Within each node, two indices are kept into the data:

an index ordered by keys and an index ordered by the evaluation function. An expand method running in each node repeatedly removes the most promising position from the local B-cube node, expands that position, and inserts its descendants into the B-cube.

The directed search algorithm that results from using a B-cube in this manner has a number of desirable properties.

- Identical positions can be converged, since they will hash to the same key.

- The hash function in combination with the balance property of the B-cube will evenly distribute positions over the processing nodes of a concurrent computer, resulting in good load balancing.

- Perhaps most importantly, no special effort is required to make the expand method concurrent. The method simply removes a position from the B-cube, expands it, and inserts the descendants into the B-cube. All of the communication and synchronization, all of the burdens of concurrency, are handled by the B-cube.

3.9   Summary

I have developed a new data structure for implementing ordered sets, the balanced cube. The balanced cube is a distributed ordered set object. It is an ordered set of data, along with operations to manipulate those data, distributed over the nodes of a concurrent computer. Operations are initiated by messages to any node. Thus, many operations may be initiated simultaneously. The balanced cube offers significantly improved concurrency over conventional data structures such as heaps, balanced trees, and B-trees.

On sequential machines, complexity is measured by instruction counts. Based on these conventional measures, the balanced cube performs as well as balanced trees or B-trees requiring $O(\log N)$ time to search, insert, or delete a record in a structure of N records. For concurrent machines, however, communications costs are more important than instruction counts, and the throughput of several operations executing in parallel is more important than the latency of a single operation. Based on this performance model, a balanced cube offers $O(\frac{N}{\log N})$ throughput as compared to $O(1)$ throughput for conventional data structures. Consider, for example, an $N = 1024$ processor concurrent computer. A conventional data structure implemented on such a machine can process only a single access per unit time. A balanced cube, on the other hand, can process over 100 accesses simultaneously.

In any concurrent system, consistency of interacting operations and deadlock avoidance are critical. The balanced cube is provably deadlock free. Each operation locks at most one non-deleted node at a time and unlocks this node before locking the next node. In the case of the merge operation, where there may be competition for access to deleted nodes, a priority scheme is used to resolve any conflicts. In the balanced cube, concurrently executing operations produce results that are consistent with a sequential execution of the same operations ordered by time of completion. This consistency is achieved by the judicious use of locking to make the completion of an operation appear instantaneous and to assure that the neighborhood of an operation is not modified between the time it commits to modifying the state of the cube and the time it completes performing the modification.

Balanced cubes and B-cubes can be used to construct concurrent applications. In many cases, such as in the directed search example of Section 3.8, no special effort is required to make an application concurrent. Many instances of the application simply insert and remove data from the balanced cube. The balanced cube data structure handles all communication and synchronization.

Chapter 4

Graph Algorithms

In this chapter I represent graphs as concurrent data structures and develop algorithms for manipulating graphs on message-passing concurrent computers. Unlike the ordered set structure examined in Chapter 3, a graph does not have a fixed set of operations defined on it. Instead, a graph serves as a framework for modeling and solving a number of combinatorial problems.

Graph data structures have been applied to a wide range of problem areas, including transportation, communications, computer aided design, and game playing. Because of their importance, graph algorithms for sequential machines have been studied in depth [38], [45], [65], [69], [97], and some work has been done on concurrent graph algorithms [104], [105], [118], [87]. However, little work has been done on algorithms for message-passing concurrent computers, and very little experimental work has been done to determine the performance of concurrent graph algorithms on large (> 100 processor) machines.

This chapter addresses these gaps in the literature by formulating new concurrent graph algorithms for three important graph problems and evaluating their performance through both analysis and experiment. Section 4.2 discusses concurrent shortest path algorithms. A weakness in an existing concurrent shortest path algorithm is exposed, and a new algorithm is developed to overcome this problem. Max-flow algorithms are discussed in Section 4.3. Two new max-flow algorithms are developed. Finally, Section 4.4 deals with the graph partitioning problem. Novel techniques are developed to prevent thrashing of vertices between partitions and to keep the partitions balanced concurrently.

4.1   Nomenclature

Definition 4.1 A *graph* $G(V, E)$ consists of a set of *vertices*, V, and a set of *edges*, $E \subseteq V \times V$. The source vertex of edge e_n is denoted s_n and the destination, d_n.

Definition 4.2 A *path* is a sequence of edges $P = e_1, \ldots, e_k \ni \forall i \ d_i = s_{i+1}$. The source of the path is $s_P = s_1$ and the destination of the path is $d_P = d_k$.

Definition 4.3 A path P is said to visit a vertex v if P contains an edge e_n and $v = s_n$ or $v = d_n$. A *proper path* visits no vertex twice.

Definition 4.4 The *degree* of a vertex, v, is the number of edges incident on v. The in-degree of v is the number of edges with destination v and the out-degree of v is the number of edges with source v.

Most graphs encountered in computer aided design and transportation problems are sparse: $O(|E|) \approx O(|V|)$. For this reason I restrict my attention to sparse graphs.

The CST headers for classes Graph, Vertex and Edge are shown in Figure 4.1. A graph is represented by two distributed collection objects, vertices, V, and edges, E. Elements of vertices are of class Vertex and consist of forward and backward adjacency lists. The adjacency list representation is used here, since it is more efficient than adjacency matrices in dealing with the sparse graphs characteristic of most large problems. Each edge in the graph is an instance of class Edge which relates its source and destination vertices.

In the following sections I will define subclasses of Vertex and Edge to include problem specific instance variables such as length, weight, capacity and flow. To conserve space, these subclasses will not be explicitly declared. Instead, the new instance variables in each subclass will be informally described.

4.2   Shortest Path Problems

The shortest path problem has wide application in the areas of transportation, communication and computer-aided design. For example, finding optimal routings for aircraft, trucks or trains is a shortest path problem as is routing phone

class	Graph	*generic graph*
superclass	Object	
instance variables	vertices	*a distributed collection*
	edges	*a distributed collection*
class variables		*none*
locks		*none*

class	Vertex	
superclass	Object	
instance variables	forwardEdges	
	backwardEdges	
class variables		*none*
locks		*none*

class	Edge	
superclass	Object	
instance variables	source	$s, \text{ where } e = (s, d)$
	dest	$d, \text{ where } e = (s, d)$
class variables		*none*
locks		*none*

Figure 4.1: Headers for Graph Classes

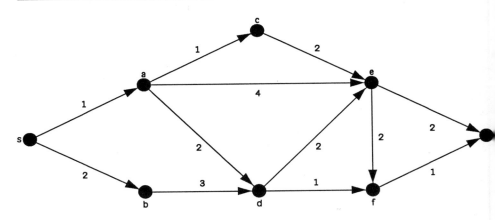

Figure 4.2: Example Single Point Shortest Path Problem

calls in a telephone network. Shortest path algorithms are also used to solve computer-aided design problems such as circuit board routing and switch level simulation [14].

To discuss *shortest* paths, we must first define length.

Definition 4.5 *Length*, l, is a function $E \rightarrow \mathcal{R}$. The length of a path is the sum of the edge lengths along the path $l(P) = \sum_{e_j \in P} l(e_j)$.

Definition 4.6 The *diameter*, D, of a graph, G, is the maximum over all pairs of points of the minimum length of a path between a pair of points,

$$D = \max \{\min l(P) | s_P = v_i, d_P = v_j\} \ \forall \ v_i, v_j \in V \tag{4.1}$$

4.2.1   Single Point Shortest Path

The *single point shortest path problem* (SPSP) involves finding the shortest path from a distinguished vertex, $s \in V$ to every other vertex. In this section I examine an existing concurrent SPSP algorithm due to Chandy and Misra [15] and show that it has exponential complexity in the worst case. I go on to develop a new concurrent algorithm for the SPSP problem that overcomes

---

```
spsp: s
   | vSet u v |
   vertices do: [:aVertex | aVertex distance: infinity].
   source distance: 0.
   vSet ←SortedCollection sortBlock:[:a :b | a distance < b distance].
   vSet add: source.
   [vSet isEmpty] whileFalse: [
       u ←vSet removeFirst.
       (u forwardEdges) do: [:edge |
           v ←edge destination.
           ((u distance + edge length) < v distance) ifTrue:[
               v distance: (u distance + edge length).
               v pred: u.
               vSet add: v]]]
```

Figure 4.3: Dijkstra's Algorithm

---

the problem of Chandy and Misra's algorithm and requires at most $O(|V|^2)$ messages.

The SPSP problem was solved for sequential computers by Dijkstra in 1959 [29]. Shown in Figure 4.3, Dijkstra's algorithm begins at the source and follows edges outward to find the distance from the source to each vertex. The *wavefront* of activity is contained in vSet, the set of vertices that have been visited but not yet expanded. To avoid traversing an edge more than once, the algorithm keeps vSet in sorted order. Each iteration through the whileFalse: loop, the active vertex nearest the source, u, is removed from vSet and expanded by updating the distance of all forward neighbors. When the algorithm terminates, the distance from source to a vertex, v, is in v distance and the path can be found by following the pred links from v back to source. Dijkstra's algorithm remains the best known algorithm for the sequential SPSP problem.

A trace of Dijkstra's Algorithm on the graph of Figure 4.2 is shown in Figure 4.4. For each iteration of the whileFalse: loop, the figure shows the vertex expanded, its distance from the source, its predecessor and the state of the active set. Note that each vertex, and thus each edge, is examined exactly once. Because of this property, for sparse graphs Dijkstra's algorithm has a time complexity

Vertex u	Distance	Pred	vSet (vertex,dist)
s	0	nil	(a,1),(b,2)
a	1	s	(b,2),(c,2),(d,3),(e,5)
b	2	s	(c,2),(d,3),(e,5)
c	2	a	(d,3),(e,4)
d	3	a	(e,4),(f,4)
e	4	d	(f,4),(g,6)
f	4	d	(g,5)
g	5	f	

Figure 4.4: Example Trace of Dijkstra's Algorithm

of $O(|V| \log |V|)$. The loop is iterated $|V|$ times and the rate-limiting step in each iteration, selecting the vertex u, can be performed in $O(\log |V|)$ time using a heap[1].

Chandy and Misra [15] have developed a concurrent version of Dijkstra's Algorithm. This algorithm is simple and elegant; however, as we will see shortly, it has a worst case time complexity of $O(2^{|V|})$. A simplified form of Chandy and Misra's algorithm is shown in Figure 4.5. While Chandy and Misra's original algorithm uses two passes to detect negative weight cycles in the graph, the simple algorithm uses only a single pass. As with Dijkstra's Algorithm, Chandy and Misra's Algorithm works by propagating distances from the source. The algorithm is initiated by sending the source a setDistance: 0 from: nil message. When a vertex receives a setDistance:from: message, with a distance smaller than its current distance, it updates its distance and sends messages to all of its successors. Every setDistance:from: message is acknowledged with an ack message to detect termination as described in [30]. When the source replies to the graph the problem is solved and the algorithm terminates. Unlike Dijkstra's algorithm, the expansion of vertices is not ordered but takes place concurrently. This is both the strength and the weakness of this algorithm.

A trace of Chandy and Misra's algorithm on the graph of Figure 4.2 is shown in Figure 4.6. Each row of the figure corresponds to one arbitrary time period. Each column corresponds to one vertex. For each time period, the mes-

---

[1] If there are only a constant number of edge lengths, then the selection can be performed in constant time using a bucket list and the time complexity of the algorithm is $O(|V|)$.

instance methods for class Path Graph
   spsp: s
      | |
       source setDistance: 0 from: nil.

instance methods for class Path Vertex
   setDistance: aDist from: aVertex
      | |
      (aDist < distance) ifTrue: [
         distance ←aDist,
         (pred notNil) ifTrue:[pred ack].
         pred ←aVertex.
         forwardEdges do: [:edge |
            (edge destination) setDistance: (distance + edge length) from: self
            nrMsgs ←nrMsgs + 1]].
        ifFalse: [aVertex ack].

   ack
      | |
      nrMsgs ←nrMsgs - 1.
      (nrMsgs = 0) ifTrue:[
         (pred notNil) ifTrue: [pred ack].
         (self = graph source) ifTrue: [graph reply].
         pred ←nil].

Figure 4.5: Simplified Version of Chandy and Misra's Concurrent SPSP Algorithm

Time	a	b	c	d	e	f	g
1	(s,1)	(s,2)					
2			(a,2)	(b,5)	(a,5)		
				(a,3)			
3					(d,7)	(d,4)	(e,7)
					(d,5)	(e,7)	
					(c,4)		
4						(e,6)	(e,6)
							(f,5)

Figure 4.6: Example Trace of Chandy and Misra's Algorithm

sages (vertex, distance) received by the vertices are shown in the corresponding columns. For instance, during the first time period vertex a receives the message setDistance: 1 from: s, or (s,1) and vertex b receives (s,2).

The order of message arrival at reconvergent vertices is nondeterministic. Figure 4.6 shows a particularly pessimistic message ordering to illuminate a problem with the algorithm. During time period 2, messages (b,5) and (a,3) are received by vertex d. In the example I assume the message from b arrives before the message from a. Vertex d updates its distance twice and sends two messages to vertex e. Unlike Dijkstra's algorithm, Chandy and Misra's algorithm may traverse an edge more than once.

This multiple edge traversal, due to the very loose synchronization of the algorithm, can result in exponential time complexity. Consider the graph of Figure 4.7. If messages arrive in the worst possible order, Chandy and Misra's algorithm requires $O(2^{\frac{|V|}{2}})$ time to solve the SPSP problem on this graph. Each triangular stage doubles the number of messages. Vertex v_1 receives messages with distances 3 and 2; v_2 receives 7,6,5 and 4; v_k receives $2^k + 2k - 1, \ldots, 2k$ in that order. Although it is unlikely that the situation will ever get this bad, the problem is clear. Tighter synchronization is required.

To solve the synchronization problem with Chandy and Misra's algorithm I have developed a new concurrent algorithm for the SPSP problem that synchronizes all active vertices. This algorithm, shown in Figure 4.8, synchronizes all vertices

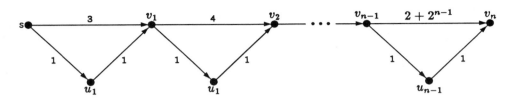

Figure 4.7: Pathological Graph for Chandy and Misra's Algorithm

in the graph with their neighbors. By forcing a vertex to examine all of its input edges before propagating a value on its output edges, the worst case time complexity of the algorithm is reduced to $O(|V|)$ for sparse graphs[2]. The worst case number of messages required for sparse graphs is $O(|V|^2)$.

The algorithm is initialized by sending an spsp: source message to the graph. The graph then initializes each non-source vertex by sending it an spspInit: ∞ message. The source receives an spspInit:0 message. The spspInit messages initialize the distance instance variable of each vertex and start the synchronized distance computation by having each vertex send setDist:from: messages to all of its forward neighbors.

Figure 4.9 illustrates the synchronization imposed by this algorithm on each vertex by means of a Petri Net [100]. During each step of the algorithm, each vertex sends setDist messages to all of its forward neighbors. When setDist messages have arrived from all backward neighbors, the vertex acknowledges these messages with ackDist messages. When ackDist messages are received from all forward neighbors, the cycle begins again. Using this mechanism, vertices are kept locally synchronized. They do not operate in lockstep, but, on the other hand, two vertices cannot be out of synchronization by more than the number of edges separating them.

The algorithm as presented will run forever since no check is made for completion. Completion detection can be added to the algorithm in one of two ways.

---

[2]On any real concurrent computer $O(|V|)$ performance will not be seen, since it ignores communication latency between vertices. On a binary n-cube processor, for example, the average latency is $O(\log N)$, where N is the number of processors, giving a time complexity of $O(|V| \log N)$.

instance methods for class Path Graph

spsp: s
    | |
    vertices do: [:vertex |
        (vertex = source) ifTrue: [vertex spspInit: 0]
        ifFalse: [vertex spspInit: ∞]]

instance methods for class Path Vertex

setDist: aDist over: anEdge
    | |
    nrMsgs ←nrMsgs - 1,
    (aDist $<$ distance) ifTrue: [
        distance ←aDist,
        pred ←(anEdge source)],
    (nrMsgs =0) ifTrue:[
        self sendAcks,
        (nrAcks = 0) ifTrue: [self sendMsgs]]

spspInit: aDist
    | |
    distance ←aDist,
    self sendMsgs

sendMsgs
    | |
    nrAcks ←(forwardEdges size),
    acksSent ←false,
    forwardEdges do: [:edge | (edge destination) setDist: (distance + edge length) over: edge]

sendAcks
    | |
    nrMsgs ←(backwardEdges size),
    acksSent ←true,
    backwardEdges do: [:edge | (edge source) ackDist]

ackDist
    | |
    nrAcks ←nrAcks - 1.
    (acksSent and: (nrAcks = 0)) ifTrue: [self sendMsgs]

Figure 4.8: Synchronized Concurrent SPSP Algorithm

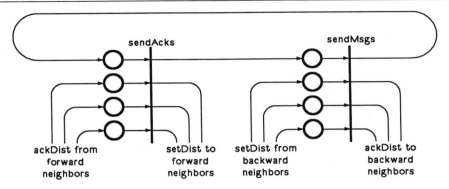

Figure 4.9: Petri Net of SPSP Synchronization

- Embed a tree into the graph. Each step, each vertex (leaf) transmits up the tree a message indicating whether or not its distance has changed. Internal nodes of the tree combine and forward these messages. When the root of the tree detects no change for h consecutive steps, where h is the height of the tree, the computation is finished.

- This shortest path is an example of a diffusing computation as defined in [30] and thus the termination detection technique described there can be applied to this algorithm.

For the sake of brevity, the details of implementing completion detection will not be described here. In the experiments described below, the second termination technique was implemented to give a fair comparison with Chandy and Misra's algorithm.

An example trace of the synchronous SPSP (SSP) algorithm on the sample graph of Figure 4.2 is shown in Figure 4.10. Since each vertex waits for distance messages on all incoming edges before propagating its next message on an outgoing edge, an unfortunate message ordering cannot cause an exponential number of messages.

Theorem 4.1 The SSP algorithm requires at most $O(|V| \times |E|)$ total messages.

Proof: In a graph with positive edge lengths, all shortest paths must be simple paths, or we could make them shorter by eliminating their cycles. Thus, a

Time	a	b	c	d	e	f	g
1	(s,1)	(s,2)	(a,∞)	(a,∞)	(a,∞)	(d,∞)	(e,∞)
				(b,∞)	(c,∞)	(e,∞)	(f,∞)
					(d,∞)		
2	(s,1)	(s,2)	(a,2)	(a,3)	(a,5)	(d,∞)	(e,∞)
				(b,5)	(c,∞)	(e,∞)	(f,∞)
					(d,∞)		
3	(s,1)	(s,2)	(a,2)	(a,3)	(a,5)	(d,4)	(e,∞)
				(b,5)	(c,4)	(e,7)	(f,∞)
					(d,5)		
4	(s,1)	(s,2)	(a,2)	(a,3)	(a,5)	(d,4)	(e,6)
				(b,5)	(c,4)	(e,6)	(f,6)
					(d,5)		

Figure 4.10: Example Trace of Simple Synchronous SPSP Algorithm

shortest path contains at most $|V| - 1$ edges. By induction we see that the algorithm finds all shortest paths containing i edges after i iterations of exchanging messages with its neighbors. Thus, at most $|V| - 1$ iterations are required. Since $|E|$ messages are sent during each iteration; $O(|V| \times |E|)$ total messages are required. ■

The experiments discussed below were performed by coding both Chandy and Misra's algorithm and the SSP algorithm in C and running them on a binary n-cube simulator. The simulator charges one unit of time for each communications channel traversed in the graph. The experiments show that for large graphs the SSP algorithm outperforms Chandy and Misra's algorithm because it has better asymptotic performance, while for small graphs Chandy and Misra's algorithm performs better since it is not burdened with synchronization overhead.

Figure 4.11 shows the speedup of both algorithms as a function of the problem size. The line marked with circles shows the speedup of Chandy and Misra's algorithm, while the line marked with diamonds shows the speedup of the SSP algorithm. The graph shows that the SSP algorithm performs better than Chandy and Misra's algorithm for large graphs.

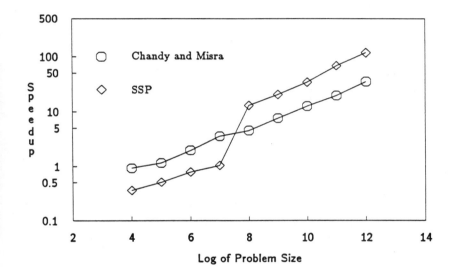

Figure 4.11: Speedup of Shortest Path Algorithms vs. Problem Size

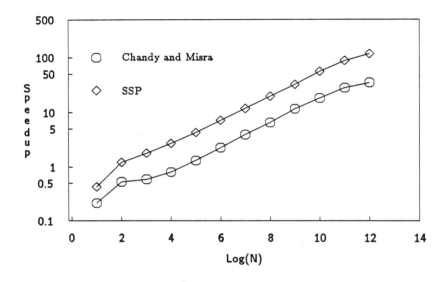

Figure 4.12: Speedup of Shortest Path Algorithms vs. Number of Processors

The algorithms were run on random graphs of degree four with uniformly distributed edge lengths. Tests were run varying the graph size in multiples of two from 16 to 4096 vertices. In each test the number of processors was equal to the number of vertices in the graph. The speedup figure in the graph is given by $\frac{T_s}{T_c}$, where T_s is the number of operations required by Dijkstra's algorithm on a sequential processor ignoring accesses to the priority queue, and T_c is the time for the concurrent algorithm on a concurrent processor. Note that these speedup figures are, in fact, pessimistic since they ignore the time required by the sequential algorithm to access the priority queue.

Figure 4.12 shows the speedup of both algorithms as a function of the number of processors. These tests were run on a random graph of degree 4 with 4096 vertices and uniformly distributed edge weights. For this graph size, the SSP algorithm is about four times as fast as Chandy and Misra's algorithm for all configurations tested. The speedup of both algorithms is $\approx \frac{N}{\log N}$ over much of

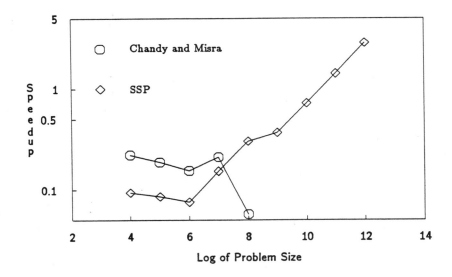

Figure 4.13: Speedup of Shortest Path Algorithms for Pathological Graph

the range with Chandy and Misra's algorithm falling short of this asymptote for large N.

Figure 4.13 shows the speedup of both algorithms for different-size instances of the pathological graph of Figure 4.7. Because the graph is very narrow and does not offer much potential for concurrency, neither algorithm performed particularly well. The SSP algorithm, however, outperformed Chandy and Misra's algorithm by a significant margin. Data are not available for Chandy and Misra's algorithm on graphs of more than 256 vertices because the algorithm did not terminate on the 512 vertex case after two days of run time on a VAX 11/750! The SSP algorithm performs moderately well even on a 4096 vertex graph.

As we will see in the next section, additional speedup can be gained exploiting concurrency at a higher level by running several shortest path problems simultaneously.

4.2.2   Multiple Point Shortest Path

In the multiple shortest path problem there are several source vertices, $s_1, \ldots, s_k$. The problem is to find the minimum length path from *each* source vertex, s_i, to every node in the graph. For example, during the loose routing phase an integrated circuit router assigns signals to channels by independently finding the shortest path from each signal's source to its destination. Since each signal is handled independently, on a concurrent computer all signals can be routed simultaneously.

The results of a number of experiments run to measure the concurrency of running multiple shortest path problems simultaneously are shown in Figures 4.14 and 4.15. Figure 4.14 shows the speedup vs. number of processors for eight simultaneous shortest path problems on graph R2.10, a random graph of degree 2 and 1024 vertices. This figure shows an almost linear speedup for small N trailing off to an $\frac{N}{\log N}$ speedup as N, the number of processors, approaches the size of the graph. This degradation is due to the uneven distribution of load that results when only a few vertices of the graph are assigned to each processing node. The maximum speedup of $\frac{N}{\log N}$ is due to the $\log N$ cost of communication in an N processor binary n-cube.

Figure 4.15 shows the speedup of the multiple path algorithm vs. the number of simultaneous problems for a fixed computer of dimension 10, 1024 nodes. For a small number of problems the speedup is limited by the number of problems available to run. As more problems are added the speedup increases to a point where it is limited by the number of processors available. Beyond this point the speedup remains at a constant level. In this experiment the processors become the limiting factor beyond 10 problems. Running a sufficient number of shortest path problems simultaneously gives a speedup that is independent of the diameter of the graph and is instead dependent on the number of available processors and the distribution of work to those processors.

The experiments shown in Figures 4.14 and 4.15 were run using Chandy and Misra's algorithm. Even greater performance gains are expected for the SSP algorithm since the multiple problems could share the significant synchronization overhead of this algorithm.

4.2.3   All Points Shortest Path

The all points shortest path problem is the extreme case of the multiple shortest path problem described above, where every vertex in the graph is a source

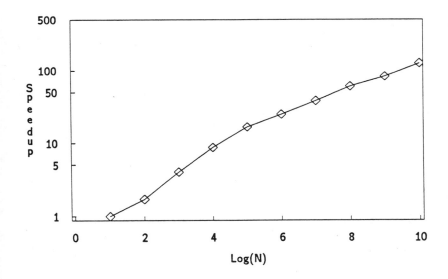

Figure 4.14: Speedup for 8 Simultaneous Problems on R2.10

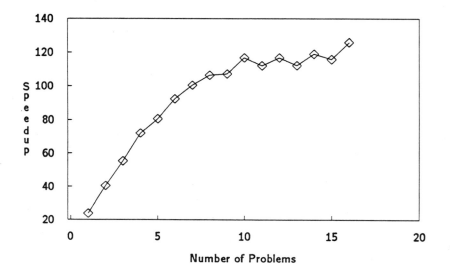

Figure 4.15: Speedup vs. Number of Problems for R2.10, n=10

floyd
```
| i j k |
  vertices do: [:vi |
     vertices do: [:vj |
        vi distTo: vj put: length of edge from i to j]]
  vertices do: [:vk |
     vertices do: [:vi |
        vertices do: [:vj |
           vi distTo: vj put: (vi distTo: vj) min: ((vi distTo: vk) + (vk distTo: vj))]]]
```

Figure 4.16: Floyd's Algorithm

vertex. An efficient sequential algorithm for solving this problem is given by Floyd [43] based on a transitive closure algorithm by Warshall [136]. This algorithm, shown in Figure 4.16, finds the shortest path between any pair of vertices, vi and vj, by incremental construction. The algorithm begins, k=0, with vi dist at: vj containing the length of the edge (if any) from vi to vj. That is, the shortest path from vi to vj containing no other vertices. On the first iteration, the algorithm considers paths from vi to vj that pass through the first vertex vk. On the mth iteration, the shortest path passing through vertices numbered less than or equal to m is found. Thus, when the algorithm completes, vi distTo: vj contains the length of the shortest path from vi to vj. This algorithm has time complexity $O(|V|^3)$ and space complexity $O(|V|^2)$.

A concurrent version of this algorithm is given in [59]. This algorithm uses $|V|^2$ processors, one for each pair of vertices, to execute the inner two loops above in a single step. $O(|V|)$ steps are required to perform the path computation. This approach is similar to one described by Levitt and Kautz for cellular automata [84]. Although it gives linear speedup, this algorithm is impractical for all but the smallest graphs because it requires $|V|^2$ processors. Since graphs of interest in computer-aided design problems often contain 10^5 to 10^6 vertices, practical algorithms must require a number of processors that grows no faster than linearly with the problem size.

Both the sequential and concurrent versions of Floyd's algorithm are very inefficient for sparse graphs. Floyd's algorithm requires $O(|V|^3)$ operations, while $|V|$ repetitions of Dijkstra's algorithm requires only $O(|V|^2 \log |V|)$ for graphs

of constant degree. This is even better than the expected case performance
of Spira's $O(|V|^2 \log^2 |V|)$ algorithm [121]. Thus, for sparse graphs, it is much
more efficient to run multiple shortest path problems as described in Section
4.2.2 than it is to run Floyd's algorithm.

The space complexity of $O(|V|^2)$ is a serious problem with the all points shortest
path problem. Note that this space requirement is inherent in the problem since
the solution is of size $|V|^2$. Another advantage of running multiple shortest path
problems instead of the all points problem is that the problem can be run in
pieces and backed up to secondary storage.

4.3   The Max-Flow Problem

The problem of determining the maximum flow in a network subject to capacity constraints, the max-flow problem, is a form of linear programming problem
that is often encountered in solving communication and transportation problems. These problems usually involve large networks and are very computation-
intensive.

Consider a directed graph G(V,E) with two distinguished vertices, the source,
s, and the sink, t. Each edge $e \in E$ has a *capacity*, $c(e)$. A *flow* function
$f : E \rightarrow \mathcal{R}$ assigns a real number $f(e)$ to each edge e subject to the constraints:

1. The flow in each edge is positive and less than the edge capacity.

$$0 \leq f(e) \leq c(e), \tag{4.2}$$

2. Except for s and t, the flow out of a vertex equals the flow into a vertex:
   vertices conserve flow.

$$\forall v \in V \setminus \{s, t\}\,, \quad \sum_{e \in \text{in}(v)} f(e) \;=\; \sum_{e \in \text{out}(v)} f(e), \tag{4.3}$$

   where in(v) is the set of edges into vertex v, and out(v) is the set of edges
   out of vertex v.

The network flow, $F(G, f)$, is the sum of the flows out of s. It is easy to show
that F is also the sum of the flows into t, the sink[3].

---

[3] It is assumed that there is no flow into the source or out of the sink.

$$F = \sum_{e\in \text{out}(s)} f(e) = \sum_{e\in \text{in}(t)} f(e) \tag{4.4}$$

The max-flow problem is to find a legal flow function, f, that maximizes the network flow F.

The max-flow problem was first formulated and solved by Ford and Fulkerson [45]. To understand their algorithm we first need the following definitions.

Definition 4.7 An edge e is useful(v,u) if either

1. $e = (v,u)$ and $f(e) < c(e)$, or

2. $e = (u,v)$ and $f(e) > 0$.

An edge that is useful(v,u) can be used to increase the flow between v and u either by increasing the flow in the forward direction or decreasing the flow in the reverse direction.

Definition 4.8 The *available flow*, a_j, over an edge $e_j = (s_j, d_j)$ is

1. $c(e_j) - f(e_j)$ in the forward direction from s_j to d_j,

2. $f(e_j)$ in the backward direction from d_j to s_j.

The available flow is the amount the flow can be increased over an edge in a given direction without violating the capacity constraint.

Definition 4.9 An *augmenting path* is a sequence of edges $e_1, \ldots, e_n$ where

1. e_1 is useful(s, v_1),

2. e_i is useful(v_{i-1}, v_i) $\forall i \ni 1 < i < n$,

3. e_n is useful(v_{n-1}, t).

Thus, an augmenting path is a sequence of edges from s to t along which the flow can be increased by increasing the flow on the forward edges and decreasing the flow on the reverse edges.

The Ford and Fulkerson algorithm begins with any feasible flow and constructs a maximal flow by adding flow along *augmenting paths*. An arbitrary search algorithm is used to find each augmenting path. Flow in each edge of the path is then increased by the minimum of the available flow for all edges in the path. The original Ford and Fulkerson algorithm may require an unbounded amount of time to solve certain pathological graphs. Edmonds and Karp [32] later discovered that restricting the search for augmenting paths to be *breadth-first* makes the time complexity of the algorithm $O(|E|^2|V|)$. For dense graphs where $|E| = O(|V|^2)$ this is quite bad, $O(|V|^5)$; however, for sparse graphs where $|E| = O(|V|)$, Edmonds and Karps's algorithm requires only $O(|V|^3)$ time. Only recently have better algorithms been discovered for sparse graphs.

Dinic introduced the use of *layering* to solve the max-flow problem [37]. Dinic's algorithm constructs a max-flow in phases. Each phase begins by constructing an auxiliary layered graph that uses only *useful* edges in the original flow graph.

Definition 4.10 A *layered graph* is a graph where the vertex set, V, has been partitioned into layers. Each vertex, v, is assigned to *layer* $l(v)$ and edges are restricted to connect adjacent layers: $\forall e = (u, v)$, $l(v) = l(u) + 1$. The layer of a vertex corresponds to the number of edges between the source and that vertex. A layered graph is constructed from a general flow graph by breadth-first search.

- The source, s, is assigned to layer $l(s) = 0$.

- For each layer i from 1 to k, a vertex u is assigned to layer i if $\exists$ an edge, e, which is useful(v, u) for some vertex v in layer $i - 1$.

During each phase of Dinic's algorithm a *maximal layered flow* is found in the layered graph using depth first search. The flows added to the layered graph are added to the original flow-graph, and the next phase of the algorithm begins by relayering the graph. The number of layers in the auxiliary graph is guaranteed to increase by one each iteration and obviously can contain no more than $|V|$ layers, so the number of iterations is at most $|V| - 1$.

Definition 4.11 A *maximal layered flow* is a legal assignment of flows to edges of a layered graph such that

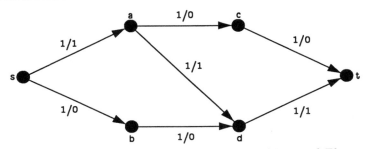

Figure 4.17: Example of Suboptimal Layered Flow

- Flows are augmented only in the forward direction. The flow over a forward edge (from layer i to layer $i + 1$) can only be increased and the flow over a reverse edge (from $i + 1$ to i) can only be decreased.

- All paths from the source to the sink are saturated.

Because of the layering constraint, a maximal layered flow is not necessarily a maximal flow on the layered network and may not be the best achievable within the constraints. For example, Figure 4.17 shows a layered graph where each edge is labeled with its capacity and flow (capacity/flow). The one unit of flow along path s, a, d, t is a maximal layered flow even though a two-unit flow is possible (paths s, a, c, t and s, b, d, t).

Finding a maximal layered flow is much easier than finding a max-flow in a general graph because each edge has been assigned a direction. Dinic's algorithm [37] constructs the layered max-flow using depth-first search which requires $O(|V| \times |E|)$ time for each phase or $O(|V|^2|E|)$ total time. Algorithms due to Karzanov [65] and Malhotra, Kumar, and Maheshwari (MKM) [86] also use layering, but construct layered max-flows by *pushing* flow from vertex to vertex. Karzanov's algorithm constructs *preflows*, pushing flow from the source, while the simpler MKM algorithm identifies a flow limiting vertex, v, and then saturates v propagating flow towards both the source and sink from v. Both of these algorithms require $O(|V|^3)$ time and Galil has shown that these bounds are tight [48]. While considerably better for dense graphs, these layered algorithms offer no improvement over Edmunds and Karp for sparse graphs.

Cherasky developed an $O(|V|^2\sqrt{|E|})$ algorithm by further partitioning the layered graph into *superlayers* [47]. Karzanov's algorithm is applied between the superlayers while Dinic's algorithm is used within each superlayer. Galil improved Cherasky's superlayer to have complexity $O(|V|^{\frac{5}{3}}|E|^{\frac{2}{3}})$ by using a set of data structures called *forests* to efficiently represent paths in the superlayers [46].

Galil and Naamad have developed an $O(|V| \times |E| \log^2 |V|)$ algorithm that uses a form of path compression. The algorithm follows the general form of Dinic's algorithm, but avoids rediscovering paths by storing *path fragments* in a 2-3 tree [47]. The fastest known algorithm for the max-flow problem, due to Sleator [120], also stores path fragments in a tree structure. Sleator's algorithm uses a novel data structure called a *biased 2-3 tree* on which join, split and splice operations can be performed very efficiently to give an $O(\log |V|)$ improvement over Galil and Naamad.

Despite the intensive research that has been performed on the max-flow problem, little work has been done on concurrent max-flow algorithms. This paucity of concurrent algorithms may be due to the fact that all of the sequential algorithms reviewed above are inherently sequential. They depend upon a strict ordering of operations and cannot be made parallel in a straightforward manner.

Shiloach and Vishkin (SV) [118] have developed a concurrent max-flow algorithm based on Karzanov's algorithm. Like Karzanov's algorithm, the SV algorithm operates in stages constructing a maximal layered flow at each stage by *pushing* preflows from the source to the sink. A novel data structure called a partial-sum tree (PS-tree) is used to make the pushing and rejection of flow efficient in dense graphs. The SV algorithm is based on a synchronized, *shared-memory* model of computation wherein all processors have access to a common memory and can even read and write the same location simultaneously. The algorithm assumes that all processors are synchronized so that all active vertices finish their flow propagation before any new active vertices begin processing. The CVF algorithm, described below, is very similar to the SV algorithm but is based on a message passing model of computation wherein shared memory and global synchronization signals are not provided.

Marberg and Gafni have developed a message passing version of the SV algorithm [87]; however, their algorithm is quite different from the CVF algorithm. The CVF algorithm is locally synchronized; vertices communicate only with their neighbors. Each *cycle* of the algorithm requires only two channel traversals for synchronization[4]. Marberg and Gafni, on the other hand, use global synchronization. All vertices are embedded in a tree which is used to broadcast *STARTPULSE* messages to all vertices to begin each cycle and to combine *ENDPULSE* messages to detect the completion of each cycle. The same tree is

---

[4]A round trip between neighboring vertices is performed each cycle.

---

maxFlow: g source: s sink: t
While an augmenting path exists from s to t
Construct a layered graph g' from graph g
Construct a maximal layered flow in g'

Figure 4.18: CAD and CVF Macro Algorithm

---

used to detect completion of each phase of the algorithm. With this approach each cycle requires a minimum of $2 \log |V|$ channel traversals for synchronization.

4.3.1 Constructing a Layered Graph

The remainder of this section describes two novel concurrent max-flow algorithms:

- the concurrent augmenting digraph (CAD) algorithm,

- the concurrent vertex flow (CVF) algorithm.

Both algorithms are similar to Dinic's algorithm in that they iteratively partition the flow-graph into layers and construct a maximal layered flow on the partitioned network. This common macro algorithm is illustrated in Figure 4.18. The algorithms differ in their approach to increasing flow in the layered network. The CAD algorithm increases flow by finding augmenting paths, while the CVF algorithm works by pushing flow between vertices.

Both the CAD and CVF algorithms construct a layered network using an algorithm similar to Chandy and Misra's shortest path algorithm. As shown in Figure 4.19, partitioning the vertices into layers is the same as finding the shortest path when all edge lengths are one.

The algorithm shown in Figure 4.19 differs from the algorithm shown in Figure 4.5 in three ways:

instance methods for class Flow Vertex
    layer: aLayer over: anEdge
        | |
      (aLayer $<$ layer) ifTrue:[
          (pred notNil) ifTrue:[pred ackFrom: self].
          layer ←layer,
          pred ←anEdge,
          forwardEdges do:[:edge |
             edge layer: (layer + 1) from: self,
             nrMsgs ←nrMsgs + 1],
          backwardEdges do:[:edge |
             edge layer: (layer + 1) from: self,
             nrMsgs ←nrMsgs +1],
          (nrMsgs = 0) ifTrue:[
             pred ackFrom: self,
             pred ←nil]]
        ifFalse:[anEdge ackFrom: self].

    ack
        | |
      nrMsgs ←nrMsgs - 1.
      (nrMsgs = 0) ifTrue:[
          (pred notNil) ifTrue:[pred ackFrom: self],
          pred ←nil].

instance methods for class Flow Edge
    layer: aLayer from: aVertex
        | |
      (aVertex = source) ifTrue:[
          (flow $<$ capacity) ifTrue:[dest layer: aLayer over: self]
          ifFalse:[aVertex ack]]
      ifFalse:[
          (flow $>$ 0) ifTrue:[source layer: aLayer over: self]
          ifFalse:[aVertex ack]].

    ackFrom: aVertex
        | |
      (self oppositeVertex: aVertex) ack

Figure 4.19: CAD and CVF Layering Algorithm

- Both forward and backward edges are used in constructing paths.

- Only edges that are useful in the proper direction are considered.

- All edge *lengths* are considered to be unity.

Restricting edge lengths to unity results in greatly improved worst case complexity. With unit edge lengths there are at most $|V|$ possible values for a vertex's distance from the source. A vertex can change its value at most $|V|$ times resulting in $O(|V|^2)$ messages in the worst case. For unit edge lengths the looser synchronization of Misra and Chandy's algorithm is preferable to the tight synchronization of the SSP algorithm. Since the algorithm performs at most $O(|V|)$ layerings in the worst case, the contribution of layering to the total number of messages required to solve the flow problem is $O(|V|^3)$.

In addition to partitioning the vertices into layers, it is also necessary to partition the edges incident on each vertex, v in layer i into a set of edges to layer $i + 1$, outEdges, a set of edges to layer $i - 1$, inEdges, and all remaining edges. Collections inEdges and outEdges will be used extensively in the following algorithms. The partitioning of edges is straightforward and will not be shown here.

4.3.2   The CAD Algorithm

The CAD algorithm constructs a maximal flow in each layered network by finding augmenting paths. Multiple paths are explored concurrently and the algorithm merges reconvergent paths into a digraph to improve performance. To prevent several paths from claiming the same edge capacity, each path is constructed in three phases: propagation, reservation and confirmation.

Propagation: All potential augmenting paths from s to t in the layered network are found by constructing a path digraph rooted at s. Construction of the path digraph begins by sending propagate messages from the source over all useful edges to vertices in layer 1. A vertex in layer i waits until it has received messages over all incoming useful edges from layer $i - 1$. It then sends propagate messages over all outgoing useful edges to layer $i + 1$. The propagation process continues until vertex t is reached.

For each edge, e, the maximum flow that can reach that edge from the source is recorded in instance variable reserveFlow. The capacity of the edges used by the paths discovered during the propagation phase is not locked, however, and several paths may use the same capacity. Conflicts over edge capacity are resolved during the reservation phase.

Reservation: Paths discovered during the propagation phase reserve edge capacity by following links in the path digraph backwards from t to s. When a propagate message reaches the sink, the reservation process is initiated by the sink sending a reserve message back to the preceding layer. A vertex in layer i waits until it receives reserve messages over all outgoing edges and then parcels the reserve flow among incoming edges. Since there may not be sufficient flow into the vertex from layer $i + 1$ to satisfy all reservations, some edges may reduce the value of reserveFlow. It is also possible that some vertices may have more incoming flow from layer $i + 1$ than can be reserved on all incoming edges. In this case the excess reservations in the higher layers will be reduced during the confirmation phase.

Confirmation: Reservations are confirmed and possibly reduced during the confirmation phase. When a reserve message reaches the source, confirmation is initiated by the source sending a confirm message back to layer 1. When a vertex in layer i has received confirm messages over all incoming edges, it partitions the flow over the outgoing edges, possibly reducing or completely canceling the reservation on some of these edges and propagates confirm messages to layer $i + 1$. Because of the way reservations are made during the reserve phase, the reservations made on incoming edges are no greater than the reservations on outgoing edges. Thus, the flow into a vertex during the confirm phase is guaranteed to be no greater than the reserved flow on outgoing edges.

The propagate methods for both vertices and edges are shown in Figure 4.20. When a non-sink vertex, v, receives a propagate message, it accumulates the total flow that could possibly reach v in instance variable inFlow and counts the number of propagate messages received in instance variable nrMsgs. When messages have been received over all incoming edges[5], a propagate message is transmitted to each outgoing edge in collection outEdges. An edge receiving a propagate message takes the minimum of the flow the vertex can deliver, aFlow, and its own available flow and propagates the resulting outFlow to the next layer of the graph. When a propagate message reaches the sink, the sink immediately sends a reserve message back to the sender to initiate the reservation phase[6].

The code that propagates reservations back toward the source is shown in Figure 4.21. A vertex, v, waits to receive reserve messages from all of its outgoing

---

[5] Recall that inEdges, outEdges is a partition of edges constructed during layering and may be different than the forwardEdges, backwardEdges partition defined by the structure of the graph.

[6] The sink could test for termination at this point by checking if any flow can reach it; however, for the sake of simplicity this test has been omitted.

instance methods for class Flow Vertex
    propagate: aFlow over: anEdge
        | |
        (isSink) ifFalse: [*internal vertex*
            inFlow ←inFlow + aFlow,
            nrMsgs ←nrMsgs + 1.
            (nrMsgs = (inEdges size)) ifTrue: [*propagate flow to next layer*
                outEdges do: [:edge | edge propagate: inFlow from: self].
                inFlow ←0,
                nrMsgs ←0]
            (outEdges size = 0) ifTrue:[*dead end, reserve 0 flow*
                inEdges do: [:edge | edge reserve: 0 from: self]]
        ifTrue: [anEdge reserve: aFlow from: self].   *sink reflects messages*
instance methods for class Flow Edge
    propagate: aFlow from: aVertex
        | outFlow |
        (aVertex = source) ifTrue:[outFlow = aFlow min: (capacity - flow)]   *forward edge*
        ifFalse:[outFlow = aFlow min: flow].   *backward edge*
        (self oppositeVertex: aVertex) propagate: outFlow over: self.
        reserveFlow ←outFlow.

Figure 4.20: Propagate Methods

instance methods for class Flow Vertex
    reserve: aFlow over: anEdge
       | outFlow |
       (isSource) ifFalse: [*internal vertex*
          inFlow ←inFlow + aFlow,
          nrMsgs ←nrMsgs + 1.
          (nrMsgs = (outEdges size)) ifTrue: [
             inEdges do: [:edge |
                outFlow ←inFlow min: edge reserveFlow.
                edge reserve: outFlow from: self,
                inFlow ←inFlow - outFlow].
             inFlow ←0,
             nrMsgs ←0]]
       ifTrue: [anEdge confirm: aFlow from: self].   *source reflects messages*
instance methods for class Flow Edge
    reserve: aFlow from: aVertex
       | |
       reserveFlow ←aFlow.
       (self oppositeVertex: aVertex) reserve: aFlow over: self.

Figure 4.21: Reserve Methods

edges, summing the reserved flow in instance variable inFlow. When v has received messages from all outgoing edges the value of inFlow represents the flow reserved between v and the sink, t. Vertex v divides this flow among its incoming vertices sending each of them a reserve message to propagate the reservations back to the next layer. A reserve message received by the source is reflected back to the sender to initiate the confirmation phase.

The details of the confirmation phase are shown in Figure 4.22. As in the propagate stage, a vertex, v, waits for messages on all incoming edges before sending messages over all outgoing edges. When v receives the confirm message from the last incoming edge, instance variable inFlow represents the amount of flow that has been added to paths from the source, s, to v. Vertex v uses this flow to confirm reservations on outgoing edges until it is used up. If the incoming flow is not sufficient to satisfy all outgoing reservations, one outgoing edge may only have part of its reservation confirmed (aFlow < reserveFlow) and some edges may have their reservation completely canceled (aFlow = 0). An edge receiving a confirm message increments or decrements its flow by the specified amount depending on whether it is a forward or backward edge.

When all confirm messages reach the sink, t, an iteration is complete and t replies with the added flow to the macro-level algorithm. If the added flow is zero, a maximal layered flow has been constructed and the macro-level algorithm proceeds to re-layer the network for the next solution phase. Otherwise, another iteration of path finding is initiated by sending a propagate message to the source.

Lemma 4.1 Each iteration of the CAD algorithm saturates at least one vertex, v, leaving no useful flow into v.

Proof: There are two cases:

1. No vertex reduces the reservation by having inFlow > 0 after sending all of its reserve messages: all flow propagated into the sink is confirmed. Proof by induction on the number of layers, l.

   - For $l = 1$, since the flow propagated into the sink comes directly from the source, confirming this flow saturates all edges into the sink and thus saturates the sink vertex.

   - Consider a network of l layers. The flow propagated along each edge is the minimum of the available flow on the edge and the maximum flow that can reach the preceding vertex. Thus, for each edge into

instance methods for class Flow Vertex
    confirm: aFlow over: anEdge
      | outFlow |
      (isSink) ifFalse: [*internal vertex*
          inFlow ←inFlow + aFlow,
          nrMsgs ←nrMsgs + 1.
          (nrMsgs = (inEdges size)) ifTrue: [
              outEdges do: [:edge |
                  outFlow ←inFlow min: edge reserveFlow.
                  edge confirm: outFlow from: self,
                  inFlow ←inFlow - outFlow].
              nrMsgs ←0]]
      ifTrue: [*sink*
          inFlow ←inFlow + aFlow,
          nrMsgs ←nrMsgs + 1.
          (nrMsgs = (inEdges size)) ifTrue: [requester reply: inFlow]].
instance methods for class Flow Edge
    confirm: aFlow from: aVertex
      | |
      reserveFlow ←0.
      (aVertex = source) ifTrue:[flow ←flow + aFlow]   *forward edge*
      ifFalse:[flow ←flow - aFlow].   *back edge*
      (self oppositeVertex: aVertex) confirm: aFlow over: self].

Figure 4.22: Confirm Methods

the sink, either that edge is saturated, or all flow propagated to the preceding vertex will be confirmed. If all edges into the sink are saturated, then the sink is saturated. If some edge $e = (v, t)$ into the sink is not saturated, then all flow propagated into vertex v is confirmed. This situation is analogous to vertex v being the sink vertex of a $l - 1$ layer network. Thus by induction, some vertex will be saturated.

2. If some vertex v reduces the reservation, then $\exists$ a vertex $u \ni u$ reduces the reservation and no vertex in a layer lower than $l(u)$ reduces the reservation. Thus, all flow propagated into u is confirmed. Consider u as the sink of a graph of depth $l(u)$; then by the result of case (1) above, some vertex in this subgraph is saturated.

■

Theorem 4.2 The CAD algorithm requires $O(|V|^2|E|)$ messages.

Proof: The CAD algorithm sends exactly $3 \times |E|$ messages during each iteration of the three phases. By Lemma 4.1 each iteration saturates at least one vertex, so there can be at most $|V|$ iterations per layering. Since at most $|V| - 1$ layerings are constructed, the total number of messages sent is at most

$$3 \times |E| \times |V| \times (|V| - 1) = O(|V|^2|E|). \tag{4.5}$$

■

4.3.3   The CVF Algorithm

Like the CAD algorithm, the CVF algorithm works by iteratively partitioning the graph into layers and then constructing a maximal layered flow for each partition. Rather than using augmenting paths to construct a maximal layered flow, however, the CVF algorithm works by pushing flow from the source vertex to the sink vertex.

The concept of a *preflow* ([65] p.53) is helpful in understanding this algorithm.

Definition 4.12 A *preflow*, $f : E \rightarrow \mathcal{R}$, is an assignment of flow to the edges of the graph so that Equation (4.2) is satisfied, but Equation (4.3) is reduced to an inequality: the flow into a vertex may exceed the flow out of a vertex.

$$\forall v \in V - \{s,t\}, \ \sum_{e \in \text{in}(v)} f(e) \leq \sum_{e \in \text{out}(v)} f(e). \tag{4.6}$$

The CVF algorithm constructs a *preflow* by pushing flow requests from source to sink. The *preflow* is converted into a maximal layered flow by rejecting excess flow requests. If a vertex, v, in layer i cannot push all requested flow on to layer $i + 1$, it rejects the remaining flow sending it back to layer $i - 1$. The vertex, u, receiving the rejected flow may send a request to another vertex in layer i, or it may reject the flow itself passing the problem back to layer $i - 2$.

This approach to constructing a maximal layered flow is not unique. The CVF algorithm is a concurrent version of Karzanov's algorithm [65]. It is almost identical to the SV algorithm [118]. There are three major differences between the CVF algorithm and the SV algorithm.

1. The SV algorithm depends on a synchronized model of computation where all vertices operate in lockstep. The CVF algorithm, on the other hand, is based on an asynchronous message-passing model of computation. Vertices operate autonomously and all synchronization is explicitly performed using message passing.

2. The SV algorithm uses PS-trees to combine communications from several edges. The CVF algorithm is intended for sparse graphs where vertex degree is small and such a structure is not needed.

3. The SV algorithm does not detect termination. All vertices become idle when a maximal layered flow has been constructed, but there is no mechanism to detect this condition. The CVF algorithm explicitly detects termination by propagating acknowledgements.

An iteration of the CVF algorithm is started by having the source vertex send request messages over all of its outgoing edges. The code for the request method is shown in Figure 4.23. When a vertex, v, in layer i has received a request message for a non-zero amount of flow, it records the flow quantum requested on a LIFO stack and accumulates the flow in instance variable inFlow. When request messages have been received over all incoming edges, instance variable inFlow represents the total *unbalanced* flow into the vertex. Method sendMessages balances this flow by either pushing it to layer $i + 1$ or rejecting it back to layer $i - 1$. After its first activation, vertex v waits for messages over all incoming edges and all outgoing edges, accumulating both flow pushed from layer $i - 1$ and flow rejected from layer $i + 1$ before calling method sendMessages to balance the flow.

instance methods for class Flow Vertex
    request: aFlow over: anEdge
        | |
        (isSink) ifFalse: [*internal vertex*
           inFlow ←inFlow + aFlow,   *accumulate flow*
           (inFlow > 0) ifTrue:[stack push: aFlow@anEdge]   *record flow quanta on stack*
           nrRequests ←nrRequests + 1.
           ((nrRequests = inEdges size) and:
           ((state = #inactive) or: (nrRejects = outEdges size))) ifTrue: [
                self sendMessages]]   *distribute accumulated flow*
        ifTrue: [anEdge ackFlow: self].   *sink acknowledges immediately*

instance methods for class Flow Edge
    request: aFlow from: aVertex
        | |
        (aVertex = source) ifTrue:[flow ←flow + aFlow]   *forward edge*
        ifFalse:[flow ←flow - aFlow],   *backward edge*
        (aFlow = self availFlow: aVertex) ifTrue:[state ←#saturated].
        (self oppositeVertex: aVertex) request: aFlow over: self,
        rejectableFlow ←rejectableFlow + aFlow.

    availFlow: aVertex
        | |
        (state = #active) ifTrue:[↑0].   *no more flow on saturated edge*
        (aVertex = source) ifTrue:[↑capacity - flow]   *forward edge*
        ifFalse: [↑flow]   *backward edge.*

Figure 4.23: request Methods for CVF Algorithm

instance methods for class Flow Vertex
    sendMessages
        | outFlow quantum |
        (inFlow > 0) ifTrue:[nrAcks ←0].   *reactivate unbalanced vertex*
        outEdges do: [:edge | *request flow from next layer*
            outFlow ←inFlow min: edge availFlow: self.
            edge request: outFlow from: self,
            inFlow ←inFlow - outFlow].
        ((inFlow = 0) and: (nrAcks = outEdges size)) ifTrue:[
            inEdges do: [:edge | *send acknowledges to previous layer*
                edge ackFlow: self]]
        ifFalse[
            (inFlow > 0) ifTrue: [
                state ←#saturated,
                [(inFlow > 0) and: (stack notEmpty)] whileTrue: [*reject flow to previous laye*
                    quantum ←stack pop.
                    outFlow ←inFlow min: quantum x.
                    quantum y reject: outFlow from: self,
                    inFlow ←inFlow - outFlow,
                    (quantum x > outFlow) ifTrue:[
                        stack push: quantum y@(quantum x - outFlow)]]].
            inEdges do: [:edge | edge sync: self],   *sync up previous layer*
            (state = #saturated) ifTrue:[state ←#active]].   *become active*
        inFlow ←0,   *reset flow rejected to source*
        nrRequests ←0,   *reset message counts*
        nrRejects ←0.

Figure 4.24: sendMessages Method for CVF Algorithm

Method sendMessages shown in Figure 4.24 balances the flow at a vertex, v, and synchronizes v with its neighbors. To balance the flow at vertex v the method first tries to push the excess flow to layer $i + 1$ by sending request messages over output edges. These request messages propagate the preflow to the next layer of the graph. If flow remains after all requests have been sent, the remaining flow is rejected back to layer $i - 1$ by sending reject messages over incoming edges. Flow is rejected in LIFO order by rejecting flow quanta popped off the stack until the excess flow has been rejected. Once the flow has been rejected, sync messages are sent to all back edges to push the rejected flow back to layer $i - 1$ and to synchronize the algorithm. Request messages are always sent to all outgoing edges and sync or ack messages to all incoming edges to keep the algorithm synchronized. Many of these messages carry zero flow.

Method sendMessages also performs completion detection by propagating acknowledgements. Sink vertices acknowledge all flow pushed into them by sending an ackFlow message back to the sending edge. When a non-sink vertex, v, receives acknowledgement from all of its neighbors in layer $i + 1$ and receives no additional flow requests, it sends acknowledgments to all of its neighbors in layer $i - 1$. These acknowledgements, however, can be canceled by sending a non-zero flow request to v. When the source receives acknowledgements from all of its neighbors, completion is detected and the algorithm terminates.

Figure 4.25 shows the details of rejection. When an edge, e, receives a reject message, it adjusts its flow accordingly and changes its state to either #saturated (no more flow can be requested across e) or #done (no more flow can be requested or rejected across e). Flow rejections are accumulated until e receives a sync message. The sync message causes e to propagate the rejected flow back to the vertex at its opposite end. Vertices handle flow rejections exactly the same as flow requests: flow is accumulated until all requests and rejections are in, and then the vertex is balanced by calling sendMessages.

Both vertices and edges have a state encoded in instance variable state. Edge states progress from #active to # saturated, and finally to #done.

#active: All edges begin each CVF iteration in the #active state. Flow can be requested only across active edges.

#saturated: When the maximum possible flow has been requested across an #active edge, or when any flow is rejected across an edge, the edge becomes #saturated. No further flow can be requested across a #saturated edge.

instance methods for class Flow Vertex

    reject: aFlow over: anEdge

      | outFlow |

      inFlow ←inFlow + aFlow,   *accumulate flow*

      nrRejects ←nrRejects + 1.

      ((nrRejects = outEdges size) and: (nrRequests = inEdges size)) ifTrue: [

          self sendMessages]].   *distribute excess flow*

    ackFlow: anEdge

      | |

      nrAcks ←nrAcks + 1.   *count acks*

      self reject: 0 over: anEdge.

instance methods for class Flow Edge

    reject: aFlow from: aVertex

      | |

      (aVertex = source) ifTrue:[flow ←flow + aFlow]   *forward edge*

      ifFalse:[flow ←flow - aFlow],   *backward edge*

      rejectableFlow ←rejectableFlow - aFlow.

      rejectedFlow ←rejectedFlow + aFlow

      (aFlow > 0) ifTrue:[

          (rejectableFlow = 0) ifTrue:[state ←#done]   *no more flow to reject*

          ifFalse:[state ←#saturated]],   *no more requests*

    sync: aVertex

      | |

      rejectableFlow ←rejectableFlow - rejectedFlow,

      (self oppositeVertex: aVertex) reject: rejectedFlow over: self,

      rejectedFlow ←0.

    ackFlow: aVertex

      | |

      (state = #saturated) ifTrue:[state ←#done].

      (self oppositeVertex: aVertex) ackFlow: self.

Figure 4.25: reject and ackFlow Methods for CVF Algorithm

#done: When all requested flow is rejected across an edge, or the flow in a #saturated edge is acknowledged, the edge becomes #done. The flow in a #done edge cannot be changed.

Vertex states progress from #inactive to #active to #saturated:

#inactive: To initiate synchronization, all vertices begin in the #inactive state. Inactive vertices wait only for messages on their incoming edges before calling sendMessages to balance their flow. After their first balancing operation, all vertices become #active or #saturated.

#active: As with edges, a vertex remains #active until it rejects flow.

#saturated: Once a vertex rejects flow, it becomes #saturated and will no longer accept flow requests.

Lemma 4.2 Each iteration of the CVF algorithm constructs a maximal layered flow.

Proof:

- The flow is legal since acknowledges are only propagated back from the sink to the source when all vertices are balanced.

- Suppose the flow was not maximal; then there exists an augmenting path, P, in the layered network. Let v_i be the vertex of P in the ith layer of the graph. The source requests all possible flow from v_1, so some vertex on P must have rejected some of this flow. Let $v_j = t$ be the vertex of P furthest from the source that rejected the flow. Each vertex v_i requests all possible flow from all of its neighbors in layer $i + 1$ including v_{i+1} before rejecting any flow to v_{i-1}. Since v_j rejected the flow and v_{j+1} didn't, all edges out of v_j including the edge (v_j, v_{j+1}) must be saturated. Then we have a contradiction since P includes (v_j, v_{j+1}), but an augmenting path cannot contain a saturated edge.

■

The CVF algorithm is synchronized by having each vertex, v, in layer i wait for messages from all of its neighbors in layers $i \pm 1$ before sending messages to layers $i \pm 1$. This synchronization, illustrated in the Petri Net of Figure 4.26, causes operation of the layers to alternate: even layers send messages to odd

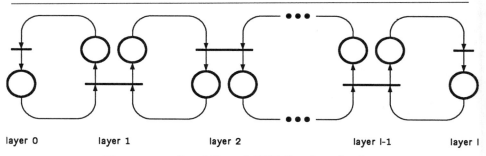

layer 0   layer 1   layer 2   layer l-1   layer l

Figure 4.26: Petri Net of CVF Synchronization

layers and then odd layers send messages to even layers. Since the layers are not completely connected, this alternation is somewhat loose; however, the Petri Net assures us that each vertex will execute the same number of message sending cycles.

Lemma 4.3 Each iteration of the CVF algorithm requires at most $O(|V|)$ cycles and thus $O(|V|^2)$ messages.

Proof: Flow pushed from the source is either acknowledged or rejected. Acknowledged flow takes $O(|V|)$ cycles to reach the sink from its last point of rejection. Rejected flow performs a depth-first search (DFS) of the layered graph before it is either rejected back to the source or is acknowledged by the sink. Flow first pushes forward (depth-first); then, if it is rejected, it follows the same path backward, since requests are rejected in a LIFO manner. Each time flow backtracks over a node, that node is saturated and will not be visited again. In the worst case a single flow quantum traverses the entire layered graph taking $O(|V|)$ cycles. Since every vertex sends messages every cycle, $O(|V|^2)$ messages are required. If several flow quanta are being rejected simultaneously, the traversal takes less time. ■

To see that this bound is tight, consider the graph of Figure 4.27, a binary tree where all internal edges have capacity 100 and all leaves are connected to the sink with capacity 1. The CVF algorithm will perform DFS on this graph taking $2|V|$ cycles to construct a maximal layered flow. In contrast, the CAD algorithm will find a maximal layered flow for this graph in $O(\log|V|)$ cycles.

The graph of Figure 4.27 illustrates the major difference between the CAD and CVF algorithms. In the CAD algorithm all potential paths from source to sink

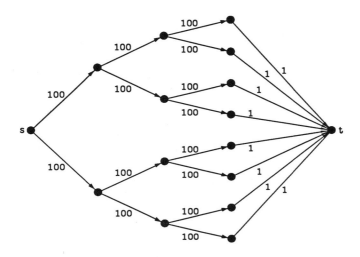

Figure 4.27: Pathological Graph for CVF Algorithm

are discovered simultaneously without considering possible conflicts. The CVF algorithm, on the other hand, never generates any conflicts. It explores only those paths that have guaranteed available capacity on their initial segments. This conservative approach to augmenting flow can result in sequential execution for graphs like the one shown in Figure 4.27 that bottleneck near their sink.

Theorem 4.3 The CVF algorithm requires at most $O(|V|^3)$ messages.

Proof: The contribution of layering is $O(|V|^3)$. By Lemma 4.2 a maximal layered flow is constructed by each iteration of the CVF algorithm. Since at most $|V| - 1$ layerings are produced, at most $O(|V|)$ iterations are performed. By Lemma 4.3 each iteration takes $O(|V|^2)$ time. Thus, the algorithm requires $O(|V|^3)$ time. ∎

4.3.4 Distributed Vertices

In both the CAD and CVF algorithms, the source and the sink are bottlenecks that serialize the algorithm. At most one path can be processed by the source

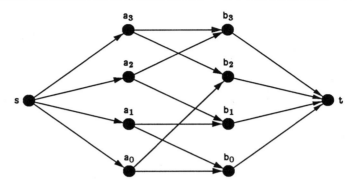

Figure 4.28: A Bipartite Flow Graph

or sink per unit time and each path must pass through both the source and sink twice. The problem is especially acute in the case of a flow-graph for solving a bipartite matching problem where the fanout of the source and fan-in of the sink are $\frac{|V|}{2} - 1$ as shown in Figure 4.28.

The source and sink bottlenecks can be removed by distributing these vertices. The only operations performed at the source and sink are keeping message counts and reflecting messages back across edges. Messages to or from the source or sink on a particular edge affect no other edges. Thus, we can split the source and sink into multiple vertices: one for each edge incident on the original source or sink as shown in Figure 4.29. The individual source and sink vertices act independently, reflecting messages and keeping message counts. When all source vertices have been acknowledged (in the CVF algorithm) or have received confirm messages (in the CAD algorithm), completion is detected and the algorithm terminates.

4.3.5   Experimental Results

The CAD and CVF algorithms have been run on a concurrent computer simulator to measure their performance experimentally. Dinic's sequential max-flow algorithm was also tested to give a performance baseline for comparisons. Randomly generated bipartite graphs with uniformly distributed edge capacities were used as test cases for the max-flow algorithms. The tests were run on a simulated binary n-cube interconnection network where one unit of time is

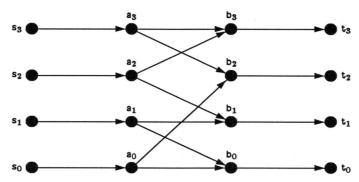

Figure 4.29: Distributed Source and Sink Vertices

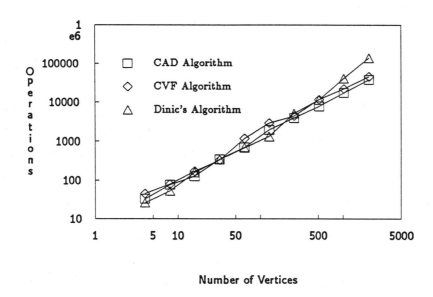

Number of Vertices

Figure 4.30: Number of Operations vs. Graph Size for Max-Flow Algorithms

charged for each channel traversed by a message. This implies that a random message requires on the average $n = \log N$ units of time to reach its destination. The results of these experiments are shown in Figures 4.30, 4.31 and 4.32.

The number of messages required by each of the algorithms as a function of graph size is shown in Figure 4.30. For purposes of comparison, Dinic's algorithm was charged one message for each edge traversed. While the worst-case complexity of these three algorithms is $O(|V|^3)$, all three give linear performance on the test cases. The CAD algorithm (squares) requires the fewest messages, $\approx 9|V|$, followed by the CVF algorithm (diamonds) with $\approx 11|V|$, and, finally, Dinic's sequential algorithm (triangles) required $\approx 30|V|$ edge traversals to construct a max-flow. This figure shows that the CAD and CVF algorithms are, in fact, good sequential algorithms. The overhead of synchronization does not greatly increase the number of messages required when compared to a strictly sequential algorithm. The CAD algorithm requires fewer messages than the CVF algorithm because it propagates wavefronts of activity across the graph. Only the vertices on the wavefront are active at a given time. In contrast, the CVF algorithm is tightly synchronized with all vertices actively sending messages all the time.

The speedup of the two concurrent algorithms relative to the sequential algorithm is shown in Figure 4.31 as a function of the number of processors for a 4096 vertex graph. Both the CAD algorithm (squares) and the CVF algorithm (diamonds) show nearly linear speedup until they saturate at 128 processors with speedups of close to 200. The speedup is greater than the number of processors because the CAD and CVF algorithms are better than Dinic's algorithm even for a single processor. The speedup of the CAD algorithm varies from 2.5 on a single processor to 204 on 256 processors, a relative speedup of 81.6. The speedup of the CVF algorithm varies from 1.7 on a single processor to 202 on 1024 processors for a relative speedup of 119. As expected the CAD algorithm performs slightly better for small numbers of processors with the CVF algorithm catching up for large numbers of processors.

Figure 4.32 shows the speedup of the CAD (squares) and CVF (diamonds) algorithms as a function of graph size. Each test was run with $\frac{|V|}{2}$ processing nodes. For the CAD algorithm, speedup varies from 0.9 for a 4 vertex graph to 196 for a 4096 vertex graph. CVF speedups were nearly the same, varying from 0.7 for a 4 vertex graph to 202 for a 4096 vertex graph. The speedup grows slower than linearly, almost logarithmically, as the number of vertices is increased from 4 to 128 and then just about linearly from 128 to 4096 vertices. This irregularity in the speedup curve may be due to the fact that only one graph of each size was tested.

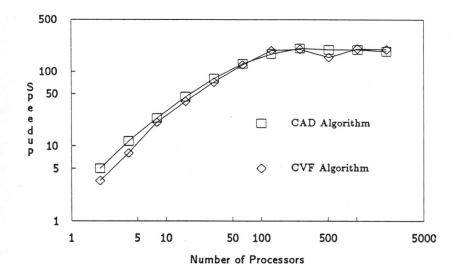

Figure 4.31: Speedup of CAD and CVF Algorithms vs. No. of Processors

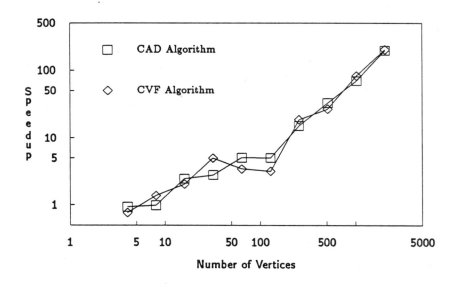

Figure 4.32: Speedup of CAD and CVF Algorithms vs. Graph Size

4.4 Graph Partitioning

The graph partitioning problem involves partitioning the vertices of a graph into two sets in a manner that minimizes the sum of the weights of edges incident on both sets. This problem has important applications in computer aided design where graphs representing the interconnection of logic circuits are partitioned onto several physical packages [107]. Graph partitioning is also used in process placement on multiprocessors where a possibly dynamic graph representing the interconnection of logical processes is partitioned over a set of physical processors [122].

Unfortunately this important problem is NP-Complete [49]. In practice, however, polynomial time heuristics based on iterative improvement methods are used with good results [69],[40].

In the Kernighan and Lin algorithm [69], an initial partition is improved by exchanging pairs of vertices between the two sets. At each step, the pair that results in the greatest reduction in the weight of the cutset is chosen for exchange. The limiting step of the algorithm is computing the weight reduction associated with each pair and sorting the pairs according to this number. Based on this step, the time complexity of the algorithm is estimated to be $O(|V|^2 \log |V|)$.

Fidducia and Mattheyses [40] improve upon this algorithm to give a linear-time heuristic. Their most important modification is to consider single vertex moves rather than pairwise exchanges. They also use a bucket list to sort the vertices so vertices can be added or deleted from the list in constant time.

A novel approach to the graph partitioning problem using linear programming has been developed by Barnes [6]. This approach converts the partitioning problem into a matrix approximation problem. The matrix approximation problem is then solved using linear programming. This method is good for finding an approximate solution near a local minimum for the problem. Barnes then uses an iterative improvement algorithm similar to that of Kernighan and Lin to fine-tune this approximate solution.

The partitioning problem can also be approximately solved using simulated annealing [71]. Simulated annealing, as applied to graph partitioning, involves randomly selecting a move to alter the partition, and then accepting this move with a probability dependent on its gain and the current annealing *temperature*. At high temperatures most moves are accepted regardless of gain. As the graph *cools*, the algorithm becomes more selective, accepting fewer negative gain moves. At zero temperature only positive gain moves are accepted. This technique generally achieves better solutions than the straight iterative improvement algorithms, because by occasionally accepting bad moves it is capable of avoiding local minima. Simulated annealing requires considerably more computing time than the other methods.

Consider an undirected graph $G = (V, E)$ where edges $(v_1, v_2) = e \in E$ are assigned weight $w(e)$. The vertices are partitioned into two disjoint sets A and B.

Definition 4.13 The *cut* defined by A and B is the set of edges $C(A, B) = \{(a, b) \mid a \in A,\ b \in B\}$. The sum of the weights of edges in the cut is the weight of the cut

$$W(A, B) = \sum_{e \in C(A,B)} w(e). \tag{4.7}$$

Definition 4.14 The imbalance of a partition A, B is

$$I(A, B) = ||A| - |B||. \tag{4.8}$$

The object of a graph partitioning algorithm is to find a partition of V into A, B, subject to a balance constraint $I(A, B) < C_b$ so as to minimize the weight of the cut, $W(A, B)$.

The remainder of this section describes a novel concurrent heuristic graph partitioning algorithm. Like the sequential algorithms described in [69] and [40], it is an iterative improvement algorithm. Starting from an initial partition, vertices are moved from one set to the other to improve the partition. The algorithm is concurrent in that it moves many vertices simultaneously while sequential algorithms move only one or two vertices at a time.

4.4.1   Why Concurrency is Hard

Concurrency introduces two major problems: thrashing and balancing. There are cases where making several simultaneous moves increases the weight of the cut even though each move taken individually would reduce the weight of the cut. The simplest example of this thrashing problem is shown in Figure 4.33. Vertices $a \in A$ and $b \in B$ are connected with weight w_h to each other, and with weight w_l to another element of the same set where $w_h > w_l$. Individually, moving a to set B or b to set A would decrease the weight of the cut by $w_h - w_l$,

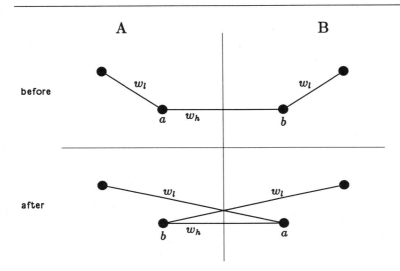

Figure 4.33: Thrashing

but moving both a and b at the same time increases the weight of the cut by $2w_l$.

A balance constraint must be imposed on the partitioning to prevent the algorithm from reducing $W(A, B)$ to zero by moving all of the vertices into one set. We require that $|I(A, B)| < C_b$ for some constant C_b. In a sequential algorithm it is quite easy to keep a running count of the size of each set. Moves are checked in sequence against the count, and only moves that keep the counts within the balance constraint are allowed. In the concurrent algorithm, this sequential checking of moves against a count is not possible, and another mechanism is required to enforce balance.

The remainder of this section develops a concurrent algorithm that meets the challenges described above. It uses a method of inhibiting gain to prevent thrashing and uses a matching tree to impose balance.

4.4.2 Gain

An iterative improvement algorithm searches a state space by applying simple transition moves to an initial state. In the case of graph partitioning, the state space is the space of all possible partitions. Each transition move is the transfer of a vertex from one set to the other. My algorithm is *greedy* in the sense that it moves all those vertices that are guaranteed to give the largest immediate *gain* in the objective function $-W(A, B)$.

Definition 4.15 The *gain* of a vertex $g(v)$ is the amount by which $W(A, B)$ is decreased by moving v from one set to another. If we define $int(v)$ to be the set of edges connecting v to vertices in the same set and $ext(v)$ to be the set of edges connecting to elements if the other set, then

$$g(v) = \sum_{e \in ext(v)} w(e) - \sum_{e \in int(v)} w(e). \tag{4.9}$$

During the first phase of the algorithm, all of the vertices compute their gain as follows.

1. All vertices transmit their set and the weight of the connecting edge to all neighboring vertices.

2. As vertices receive messages from their neighbors, they compute their gain as the sum of the weights received from vertices in the opposite set less the sum of the weights received from vertices in the same set.

4.4.3   Coordinating Simultaneous Moves

Because of the thrashing problem, if vertices were moved on the basis of gain alone, moves could potentially increase $W(A, B)$ as shown in Figure 4.34. The vertices adjacent to vertex a can be divided into four sets:

- A_m, vertices in set A with positive gain,

- A_s, vertices in set A with negative or zero gain,

- B_m, vertices in set B with positive gain,

- B_s, vertices in set B with negative or zero gain.

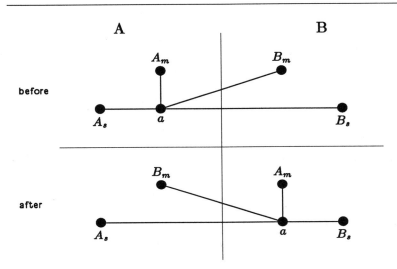

Figure 4.34: Simultaneous Move That Increases Cut

The gain of a before moving any vertices is

$$g(a) = (w(B_m) + w(B_s)) - (w(A_m) + w(A_s)). \tag{4.10}$$

Where $w(S)$ denotes the weight of all edges connecting a to set S. If all vertices with positive gain including a are moved simultaneously, the new gain of a (pushing a back into set A) becomes

$$g'(a) = (w(B_m) + w(A_s)) - (w(A_m) + w(B_s)). \tag{4.11}$$

If $w(A_s) > w(B_s)$ moving a increases the value of the cut, $W(A, B)$.

To solve this problem of simultaneously moving vertices, we inhibit vertices from moving if they are adjacent to vertices of larger gain in the opposite set. Thus, any vertex, $a \in A$, that moves knows that all of its neighbors in set B will remain stationary. The set B_m is empty and the gain, $g(a)$, is guaranteed. If some neighbor of a, $a' \in A$ moves with a to set B, the actual gain will be larger than $g(a)$. To prevent ties, two vertices a and b with equal gains $g(a) = g(b)$

compare their vertex IDs. The vertex with the larger ID inhibits the other vertex.

Inhibiting nodes from moving based on gain has the disadvantage of reducing the concurrency of the partitioning algorithm. To calculate the degradation in concurrency, assume that all vertices have degree d, and that positive gains are uniformly distributed over some range $(0, n]$. Then the probability of a vertex with gain g moving is $P_g = (\frac{g}{n})^d$. Thus, the fraction of nodes with positive gain that can be expected to move is given by

$$
\begin{aligned}
f_m &= \sum_{k=1}^{n} \frac{1}{n} P_k \\
&= \frac{1}{n} \sum_{k=1}^{n} \left(\frac{k}{n}\right)^d \\
&\leq \frac{1}{n} \int_{k=0}^{n} \left(\frac{k}{n}\right)^d dk \\
&= \int_{x=0}^{1} x^d dx \\
&= \tfrac{1}{d+1}.
\end{aligned}
\tag{4.12}
$$

Even with gain inhibition, vertices must be locked after they are moved to avoid thrashing. This is even true of sequential algorithms. To implement locking, the algorithm is performed in phases. At the beginning of each phase, all vertices are unlocked. Whenever a vertex is moved it is locked and cannot be moved again until the next phase. Phases are repeated until there are no vertices with positive gain.

Using locking and gain inhibition to prevent thrashing, the algorithm becomes

1. Set all vertices unlocked.

2. While there is some unlocked positive gain vertex,

    (a) All vertices transmit their set and the weight of the connecting edge to all neighboring vertices.

    (b) As vertices receive messages from their neighbors, they compute their gain as the sum of the weights received from vertices in the opposite set less the sum of the weights received from vertices in the same set.

(c) Unlocked vertices transmit their gain to all their neighbors. Locked vertices transmit $-\infty$ to all their neighbors.

(d) Vertices that have a positive gain greater than the gain of all of their neighbors move to the opposite set and become locked. Vertex IDs are used to break ties.

3. Repeat steps 1 and 2 until there are no positive gain vertices.

4.4.4 Balance

A balance constraint is required to prevent the algorithm from finding a cut of weight zero by moving all vertices into one set. Specifically, no move is allowed that will make one set larger than the other by more than some constant C_b. Thus, given a legal initial partition, the condition $||A| - |B|| < C_b$ will always be true.

The algorithm enforces the balancing constraint using a matching tree, a binary tree with all vertices at its leaves. At the end of the gain exchange, step 2(c) above, all vertices transmit their intentions (move or stay put) and their set (A or B) up the matching tree. Each internal vertex of the matching tree waits for all of its children to respond. It then attempts to match requests to move from set A with requests to move from set B. Matched requests are granted and the grant message is transmitted back down the tree. Unmatched requests are collected into a single message (count and set) that is transmitted up to the next level of the tree. At the root of the tree, a count of the current imbalance, $|A| - |B|$, is kept. The root acknowledges unmatched requests as long as $I(A, B) < C_b$.

With balancing, the concurrent partitioning algorithm becomes

1. Set all vertices unlocked.

2. While there is some unlocked positive gain vertex not blocked by the balancing constraint,

   (a) All vertices transmit their set and the weight of the connecting edge to all neighboring vertices.

   (b) As vertices receive messages from their neighbors, they compute their gain as the sum of the weights received from vertices in the opposite set less the sum of the weights received from vertices in the same set.

(c) Unlocked vertices transmit their gain to all their neighbors. Locked vertices transmit $-\infty$ to all their neighbors.

(d) Vertices that have a positive gain greater than the gain of all of their neighbors transmit their set and their intention to move to their parent in the matching tree. All other vertices transmit their intention to remain stationary and their set to their matching tree parent. Vertex IDs are used to break ties in gain comparison.

(e) The matching tree validates requests to move against the balance constraint as follows:

   i. Once each matching tree vertex receives messages from all of its children, it matches requests from sets A and B. Matched requests are acknowledged.

   ii. Unmatched requests are collected into a message to the next level of the matching tree.

   iii. The root of the matching tree acknowledges up to $C_b - I(A, B)$ unmatched requests from set A or up to $C_b + I(A, B)$ unmatched requests from set B and updates $I(A, B)$ accordingly. All remaining unmatched requests are rejected. If $|I(A, B)| = C_b$, all vertices in the smaller set are temporarily locked until $|I(A, B)| = C_b$.

(f) Vertices that receive acknowledgements to their requests to move become members of the opposite set.

3. Repeat steps 1 and 2 until there are no unblocked positive gain vertices.

4.4.5   Allowing Negative Moves

There are cases where any single move increases $W(A, B)$, but a sequence of moves can decrease $W(A, B)$. Consider for example the case where $|A| = |B| + C_b$, and $\forall\, b \in B$, $g(b) < 0$. There are no unblocked positive gain vertices to move. Moving a vertex b with small negative gain from B to A, however, may enable a vertex with large positive gain to move from A to B.

The algorithm can be extended to find some of these sequences by maintaining two partitions and accepting negative gain moves. The partition A', B' is updated every move and its cut weight, $W(A', B')$ is computed. The best partition A, B is updated whenever $W(A', B') < W(A, B)$.

4.4.6 Performance

Each iteration of the algorithm takes $O(d + \log |V|)$ time. Exchanging edge weights and gains between neighbors takes $O(d)$ time while propagating comparisons up the match tree takes $O(\log |V|)$ time. Since probabilistically $\frac{1}{d+1}$ of the positive gain vertices are moved in each iteration, the algorithm should complete after $O(d)$ iterations. Thus, the time complexity of the algorithm on a computer with an processor for each vertex of the graph is estimated to be $O(d^2 + d \log |V|)$ or, if we assume d is constant, $O(\log |V|)$, a speedup of $\frac{|V|}{\log |V|}$ over the linear-time sequential algorithm of Fidducia and Mattheyses [40].

4.4.7 Experimental Results

The speedup of the concurrent graph partitioning algorithm compared to a sequential algorithm similar to Fidducia and Mattheyses is shown in Figure 4.35. The tests were run on random graphs with average degree 4 and uniformly distributed edge weights. In each test the number of processors was equal to the number of vertices.

The speedup is quite disappointing for small graphs but increases significantly for large graphs. This behavior is due to the fact that the time required to perform an iteration increases very slowly with the graph size, while the number of vertices moved each iteration grows almost linearly with graph size.

The data in Figure 4.35 suggest that the efficiency of the algorithm could be improved by using fewer processors than vertices and performing balancing for all vertices on a single processor locally. This would reduce the height of the balancing tree and thus reduce the time required for each iteration of the algorithm.

For each data point shown in Figure 4.35 the concurrent and sequential algorithms produced partitions of similar weight. A partition of the same graphs performed using simulated annealing consistently produced partitions with 20% lower weight. While the gradient-following algorithms, both sequential and concurrent, get stuck in a local minimum, the simulated annealing program is able to find a point near the global minimum.

The techniques developed in this section, using gain inhibition to prevent thrashing and using a matching tree to enforce balance constraints, are completely applicable to a partitioning program that uses simulated annealing. In a concurrent simulated annealing program each vertex would compute its *inhibited*

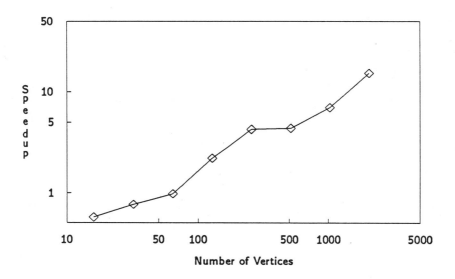

Figure 4.35: Speedup of Concurrent Graph Partitioning Algorithm vs. Graph Size

gain, the difference between its gain and the largest of its neighbors' gains. Vertices then move with a probability that is a function of their inhibited gain and the current annealing temperature. The matching tree is used to keep track of balance and to broadcast the current imbalance $I(A, B)$ to each vertex so that balance information can be incorporated in the gain function.

4.5 Summary

In this chapter I have developed concurrent algorithms for three graph problems.

In Section 4.2 I developed a new algorithm for the single point shortest path problem. Chandy and Misra's shortest path algorithm [15], because it is under-synchronized, has an exponential worst case time complexity. By adding synchronization to this algorithm I developed the SSP algorithm which has polynomial worst case time complexity. Experimental comparison of these algorithms verified that the SSP algorithm outperforms Chandy and Misra's algorithm for large graphs. Further experiments showed that additional concurrency can be attained by running several problems simultaneously. Running multiple problems is particularly advantageous for the SSP algorithm where the multiple problem instances can share the considerable synchronization overhead.

Two new algorithms for solving the max-flow problem were developed in Section 4.3. Both of the algorithms operate by repeatedly *layering* the graph and constructing a maximal layered flow. The CAD (concurrent augmenting digraph) algorithm constructs a layered flow by simultaneously finding all possible augmenting paths. These paths compete with one another for shared edge capacity through a three-step reservation process. The CVF (concurrent vertex flow) algorithm is similar to an existing concurrent max-flow algorithm [118], [87], but introduces new methods for synchronization and completion detection. Experimental results show that both of these new algorithms achieve significant speedups.

Finally, in Section 4.4 I developed a concurrent algorithm for graph partitioning. Concurrent graph partitioning is difficult for two reasons. First, moving several vertices between partitions simultaneously can result in thrashing: two vertices in opposite sets that are attracted to each other may indefinitely swap sets. Second, multiple simultaneous moves may result in a loss of balance: all vertices could simultaneously jump into the same set. The new algorithm solves the thrashing problem by using gain to inhibit simultaneous moves that might interfere with one another. The balancing problem is solved by embedding a tree into the graph. The tree matches moves in one direction with moves in the other direction to assure that the moves made during one iteration of the algorithm will not unbalance the partition.

The algorithms developed in this chapter have a great deal in common:

- They are synchronized by passing messages.

- Messages are short, containing between zero and three arguments.

- Methods are short; most are under 10 lines.

In Chapter 5 I will investigate how to build hardware to efficiently execute programs having these characteristics.

Chapter 5

Architecture

The objective of computer architecture is to organize a computer system to apply available technology to the solution of specific problems. At the Processor-Memory-Switch (PMS) level [119], architecture involves the organization of processing elements and communication channels into a computer system. At the Register Transfer (RT) level [62], architecture involves organizing registers, arithmetic units, finite state machines, and transmission lines into the processing elements and communication channels that form the building blocks at the PMS level. This chapter addresses both the RT and PMS levels of architecture.

Computer architecture cannot ignore the physical organization of the machine. VLSI computing systems are wire-limited; the complexity of what can be constructed is limited by wire density, the speed at which a machine can run is limited by wire delay, and the majority of power consumed by a machine is used to drive wires. Thus, machines must be organized both logically and physically to keep wires short by exploiting locality wherever possible. The VLSI architect must organize a computing system so that its form (physical organization) fits its function (logical organization).

I start this chapter with an intended application – the model of computation developed in Chapter 2 and the algorithms developed in Chapters 3 and 4 – and a technology – VLSI. From this starting point I develop a new architecture that takes advantage of the cost performance characteristics of VLSI technology and includes many features designed to enhance the performance of concurrent data structures.

In Section 5.1 I analyze the algorithms developed in Chapters 3 and 4. These algorithms are characterized by short messages, short methods, and a limited number of pending messages. In Section 5.3.1 I use the characteristics of these concurrent algorithms to analyze the performance of interconnection networks.

In Section 5.2 I look at VLSI technology. VLSI technology is wire-limited both by the maximum wire density that the technology can support and, since driving capacitive wires dissipates power, by the maximum power density that can be tolerated. Propagation delays in VLSI systems are also wire-limited. The delay of very short wires scales logarithmically with wire length until a critical length is reached [1]. Beyond this critical length, wire delay is bounded by the speed of light and grows linearly with wire length. In Section 5.3.1 I use these characteristics of the technology to derive some surprising results on network topology.

The development of an architecture that applies VLSI technology to support concurrent data structures is approached in two steps.

- First I consider the interconnection network over which processing elements (PEs) communicate. Based on measurements of programs and characteristics of the technology, in Section 5.3.1 I show that a 2-dimensional torus or grid network topology is preferable to a higher dimensional network. Experimental results back up this surprising result.

  In addition to a topology a network requires a routing algorithm. In Section 5.3.2 I go on to develop a new method for constructing deadlock-free routing algorithms in concurrent computer interconnection networks and apply this method to the two-dimensional torus network. The design of a self-timed VLSI chip that implements this algorithm is discussed in Section 5.3.3.

- To take advantage of a low latency communications network, the PEs must be designed to operate efficiently in the message-passing environment. In Section 5.4 mechanisms are developed to implement the model of computation described in Chapter 2 in hardware. The arrival of a message at a node results in the PE's performing the required action with a minimum of delay. Also, the sending of a message is made indistinguishable from a method call.

  To take advantage of VLSI technology, we must both exploit locality and build hardware that is specialized to particular applications. In Section 5.5 I introduce the concept of an *object expert* (OE) to achieve both of these goals. OEs exploit locality by storing objects of a particular class near the logic that operates on that class.

---

[1] This critical length is about 30mm for a typical 1.25μ CMOS technology.

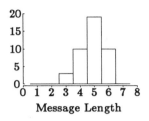 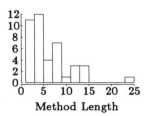

Figure 5.1: Distribution of Message and Method Lengths

5.1   Characteristics of Concurrent Algorithms

In Chapters 3 and 4, 42 CST methods were written. Here we examine these methods to find the average message length, the average method length, and the average number of pending messages per object.

Message Length

Message Length	3	4	5	6
Number of Messages	3	10	19	10

Every method has at least three fields: receiver, selector, and an implicit *reply-to* field (either the sender or the requester). Thus, the minimum message length is 3 fields; any message arguments add to this minimum length. The table above gives the static[2] frequency of message lengths for the 42 methods examined. These data are also shown in the left half of Figure 5.1. The average message length, L, is 4.9 fields. If we assume a 32-bit field size, $L \approx 160$ bits.

Method Length

Method Length	1	2	3	4	5	6	7	8	9	10	11	12	13	14	...	23
Number of Methods	4	7	6	6	3	1	3	4	0	1	2	1	2	1	0	1

[2]Static frequency is a measure of how often an event occurs in the program text. Dynamic frequency, on the other hand, is a measure of how often an event occurs during execution of the program.

CST methods tend to be quite short. The lengths of the 42 methods presented in Chapters 3 and 4 are tabulated above and shown in the right side of Figure 5.1. The average method length is 5.7 lines. While counting static method length does not account for time taken in loops, this inaccuracy is partially offset by the fact that many of the methods considered involve multiple *actions*. Because methods are short, each message received results in only a small amount of computation. Thus, the latency of message transmission must be kept very small, or excessive time will be spent transmitting messages between processing nodes and little time will be spent computing at each node.

Pending Messages

A CST object usually has only a small number of messages *pending* at any instant in time. An object typically transmits a number of messages (usually < 3) and then waits for replies from these messages before transmitting additional messages. Thus, the total number of messages in the network at any given time is a small multiple of the number of objects.

The characteristics of concurrent programs described in this section guide the development of a concurrent computer architecture in the remainder of this chapter. The message length is an important factor in deciding on the topology of the network, as described in Section 5.3.1. The short method length means that network latency is a critical parameter. Since the computation initiated by the arrival of a message takes only a short period of time, message delivery must be made fast, or all processing elements will become idle waiting for messages. Also, processing elements must be able to handle messages quickly, since the time (T_{node}) required to send a message and to initiate an action upon receipt of a message contributes to the total message latency. Finally, since each object typically has only a few messages pending at once, the required network throughput can be calculated as a function of the number of objects managed by each processing element.

Before we begin developing our concurrent architecture, we must first examine the available technology.

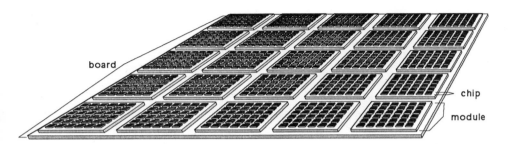

Figure 5.2: Packaging Levels

5.2 Technology

5.2.1 Wiring Density

VLSI systems (VLSI chips packaged together on modules and boards) are limited by wire density, not by terminal or logic density. Current packaging technology allows us to make more connections from VLSI chips to modules and boards than can be routed away from the chips. Since VLSI systems are wire-limited, the techniques of VLSI complexity theory [129] used to calculate bounds on the performance of VLSI chips are applicable to systems as well. In Section 5.3.1 I use this wire-cost model of VLSI systems to derive some results on concurrent computer interconnection networks.

VLSI complexity theorists, by considering the wire-limited nature of VLSI chips, have been able to prove lower bounds on the area times time squared (AT^2) required to perform a computation [89], [129]. The bound is calculated by finding the minimum *bisection width* of all possible communication graphs for the computation. Thompson shows that the area, A, of a VLSI chip is proportional to the square of the bisection width, while the time required for the computation, T, is inversely proportional to the bisection width. Thus, the quantity AT^2 is a bound independent of bisection width. By considering the wire density, not the logic density, as the limiting factor of the technology, VLSI complexity theory has been able to compute new bounds on the complexity of sorting [131], computing Fourier transforms [130], and numerous other transitive computations.

Modern high performance computers are packaged in three primary levels as shown in Figure 5.2 [117].

Chip: Circuit components and local interconnections are fabricated on a monolithic silicon die.

Module: Silicon dice are bonded to a (usually ceramic) module which provides interconnections between the chips and from the chips to board pins. Connections from chip to module can be made either by wire bonds or by solder bumps. With wire bonding the chip is placed face-up on the module, and bonds are made by running wire from pads on the periphery of the chip to corresponding pads on the module. Connections are limited to one or two rows of pads about the periphery of the chip. Typical pad dimensions are 100μ on 200μ centers. Solder bump connections are made by depositing solder bumps on the face of the chip and then placing the chip face down on the module and reflowing the solder. Solder bumps can be distributed over the face of the chip on 250μ centers [12].

Board: A number of modules are assembled on a printed circuit board (PCB) that provides interconnection between modules. Modules are connected to a PCB either by pins brazed to the back of the module that fit into holes drilled through the PCB or by surface mounting the module to the PCB in a manner similar to solder bumping chips to a module. Boards are connected using cables or backplanes.

There are often two secondary levels of packaging as well. Boards are packaged together in chassis, and chassis are assembled into racks.

Dimension	Level of Packaging			Units
	Chip	Module	Board	
Wire Width	1.25	100	200	μ
Via Diameter	1.25	100	500	μ
Wire-Hole Pitch	3.75	200	750	μ
Signal Layers	2	> 10	> 10	
Linear Size	10	100	600	mm

The table above compares the characteristics of these three levels of packaging. The PCB data are derived from design rules for a circuit board with 8 mil wire width, 8 mil spacing and 20 mil minimum hole diameter. The module characteristics are derived from available data on IBM's thermal conduction module

(TCM) [12] and a comparable ceramic technology available from Kyocera [79]. The design rules for several 1.25μ CMOS processes were consulted to construct the chip column of the table.

These numbers are for technologies that are in production today (1986). Integrated circuit design rules are halved every 4 to 6 years [93], so that by 1990 it is reasonable to expect chips to have 0.5μ wide wires. Module and PCB technologies also scale with time but at a slower rate, so that the density gap between chips ($\approx 250\frac{\text{lines}}{\text{mm}}$) and modules ($5\frac{\text{lines}}{\text{mm}}$) will continue to widen. Wafer-scale integration [99] attempts to close this gap by increasing chip size to the module level.

Most of the complexity of a VLSI system is at the chip level. Modern chips contain ≈ 2500 wiring tracks ($\frac{10\text{mm}}{3.75\mu}$), compared to 500 ($\frac{100\text{mm}}{200\mu}$) for modules and 800 ($\frac{600\text{mm}}{750\mu}$) for PCBs. While modules and PCBs can have more layers than chips, the use of additional layers is limited by the fact that in most PCB technologies, every via penetrates through the entire thickness of the board. Because chips are 50 times as dense and significantly more complex than modules, the amount of information that can be transferred from chip to module is a bottleneck that limits the performance achievable by a VLSI system.

The number of connections from a chip to a module is limited by the wiring density of the module, not, as many believe, by the number of terminals that can be placed on a chip. Consider a 10mm chip with bond pads on 250μ centers (the spacing of TCM bond pads [12]). The chip could make over 1600 connections if it were completely covered with pads. There would be no point, however, in having this number of connections. A 10mm slice of the ceramic substrate is capable of handling only ≈ 25 wires per layer[3]. Even if 10 wiring layers were used, only 250 wires could be routed away from the chip[4]. At the PCB level, assuming 10 layers and two wires between pins on a 2.5mm grid, only 80 wires can be routed out from under the chip. Even wire-bonding can achieve terminal densities that can saturate module technology. Two rows of pads on 200μ centers about the periphery of the chip would be sufficient to make 400 connections.

Wires, not terminals or logic, are the limiting factor in high-performance VLSI systems.

---

[3] Assume alternate wiring channels are used by vias to lower layers.

[4] To convert pin density to wire density, this calculation assumes that all the pins are routed to one edge of the chip.

5.2.2   Switching Dynamics

The intrinsic delay of an MOS device is the transit time, τ, the time required for a charge to cross the channel [88].

$$\tau = \frac{L}{\mu E}, \tag{5.1}$$

where μ is the carrier mobility, L is the channel length, and E is the electric field. Since $E = \frac{V}{L}$ (5.1) can be rewritten as

$$\tau = \frac{L^2}{\mu V}. \tag{5.2}$$

τ also represents the time required for a device to transfer the amount of charge on the gate, Q_g, from the drain to the source so, $i_{DS} = \frac{Q_g}{\tau}$ [52].

A more useful time measure is the delay of an inverter driving another inverter of the same size [113].

$$\tau_{\text{inv}} = \frac{C_{\text{inv}}}{C_g}\tau \tag{5.3}$$

C_{inv} is the input capacitance of the inverter, and C_g is the gate capacitance of the inverter's n-channel transistor. For a CMOS inverter with the p-channel device twice the size of the n-channel device, $\tau_{\text{inv}} = 3\tau$. In a typical 1.25μ CMOS technology with a 2.5V supply [5] voltage, $\tau = 25$ps and $\tau_{\text{inv}} = 75$ps [6].

An inverter driving a load with capacitance C_L has delay,

$$t = \frac{C_L}{C_{\text{inv}}}\tau_{\text{inv}}. \tag{5.4}$$

---

[5] As geometries get smaller, carrier velocity saturation limits device current, so that increasing the applied voltage does not reduce τ linearly. For (5.1) to hold, voltages must be scaled to keep $E < 2\frac{V}{\mu}$.

[6] Parasitic output and wiring capacitance is typically at least twice the inverter input capacitance, $C_p > 2C_{\text{inv}}$. These parasitics increase the delay of a 1.25μ inverter with a fan-out of one to ≈ 225ps.

To drive large capacitances the delay can be minimized by using an *exponential horn*, a chain of inverters with each stage e times the size of the preceding stage [88]. Using this technique, the minimum delay to drive a load from an minimum size inverter is

$$t_{\min} = \tau_{\text{inv}} e \log_e \frac{C_L}{C_{\text{inv}}}. \tag{5.5}$$

For short wires, wire delay depends logarithmically on wire length, l_W,

$$t_{\text{wire}} = \tau_{\text{inv}} e \log_e K l_W, \tag{5.6}$$

where $K = \frac{C_w}{C_g}$ and C_w is the capacitance per unit length of wire. Typically, K is in the range $0.1 < K < 0.2$. Long wires, on the other hand, act as transmission lines and are limited by the speed of light. Let l_c be the critical length at which speed of light limits transmission time. The delay of an optimally sized exponential horn driving a transmission line is the logarithmic delay of the first $\log_e K l_c - 1$ stages of the driver plus the linear delay of the wire,

$$t_{\text{longwire}} = \tau_{\text{inv}} e \left(\log_e K l_c - 1\right) + \frac{l_W \sqrt{\epsilon_r}}{c}, \tag{5.7}$$

or asymptotically,

$$t_{\text{longwire}} > \frac{l_W \sqrt{\epsilon_r}}{c}. \tag{5.8}$$

The crossover from a capacitive (short) wire to a transmission line (long) wire occurs when the delay of the last driver stage equals the time of flight along the wire, $\tau_{\text{inv}} e = \frac{l_c \sqrt{\epsilon_r}}{c}$. This equation can be rewritten as

$$l_c = \tau_{\text{inv}} e \frac{c}{\sqrt{\epsilon_r}}. \tag{5.9}$$

With $\tau_{\text{inv}} = 75\text{ps}$ and $\epsilon_r \approx 4$, the crossover from a capacitive (short) wire to a transmission line (long) wire occurs at $l_W \approx 30\text{mm}$. Thus for today's technology (1.25μ), even relatively short wires are speed-of-light limited. In an 0.5μ, technology $\tau_{\text{inv}} = 30\text{ps}$, and the crossover is at $l_W \approx 10\text{mm}$, about the length of a chip.

These speed-of-light wires are off-chip wires. As shown in Appendix C, the high resistivity of on-chip wires limits on-chip signal velocity to $\approx 8 \times 10^6 \frac{m}{sec}$.

The delay of global wires in VLSI systems is due to speed-of-light delay in the wire (not the RC delay of the driver) and thus increases linearly with wire length. For short wires, $l_W < \tau_{\text{inv}} e \frac{c}{\sqrt{\epsilon_r}}$, delay is due to the RC delay of the driver and thus grows logarithmically with wire length. In Section 5.3.1 I consider both linear and logarithmic delay models.

5.2.3 Energetics

The energy dissipated by a switching event in a VLSI system, E_{sw}, is almost entirely used to charge the capacitance of the circuit node being switched.

$$E_{\text{sw}} = \frac{1}{2}CV^2 \tag{5.10}$$

When C is the gate capacitance of a minimum-sized inverter, C_{inv}, E_{sw} is the switching energy of the technology, a figure of merit commonly used to compare logic technologies. Since V and C_{inv} both scale linearly with linear dimensions, λ, the switching energy of MOS logic scales as the cube of the linear dimensions, $E_{\text{sw}} \propto \frac{1}{\lambda^3}$.

In most VLSI systems the wiring capacitance dominates device gate capacitance, and most of the switching energy is used to drive wires. The power required to drive these wires must be supplied to each logic circuit by a power distribution system. This power, in the form of heat, must also be removed by a cooling system. The power density that the power supply and cooling systems can handle limits the performance of VLSI systems. With very advanced cooling technology [12], power densities of $30\frac{kW}{m^2}$ have been achieved.

If C_A is the capacitance per unit area and T_{cy} is the cycle time of the system, power density, P_A, is given by

$$P_A = \frac{C_A V^2}{2T_{\text{cy}}}. \tag{5.11}$$

Power density remains constant since, as voltage scales down, delay also scales down and capacitance per unit area scales up (all linearly with λ).

Consider a typical 1.25μ technology. Let us make the following assumptions:

- C_A is the capacitance of one metal layer, $C_A = 10^{-4}\frac{F}{m^2}$.

- The cycle time is 100 inverter delays, $T_{cy} = 100\tau_{inv} = 7.5\text{ns}$.

- The supply voltage, V, is 2.5V.

Then the power density is $P_A \approx 40\frac{kW}{m^2}$. Even with a very modest cycle time, the power density of a VLSI chip exceeds the capability of state-of-the-art cooling technology. Thus, power density limits the wiring density of a VLSI system independent of the wire density of the interconnection technology. We cannot escape from the problem of wiring density by adding more wire layers.

To reduce power density we must run our system more slowly. From (5.11) one would expect that power density varies as the inverse of cycle time; however, using *hot-clock*[7] logic [116], the power density can be made to scale as the inverse square of the cycle time, $P_A \propto T_{cy}^{-2}$. This relation is a strong argument for concurrency. Concurrent computing is energy efficient. We can run two computers at half speed with half the energy required to run one computer at full speed.

5.3 Concurrent Computer Interconnection Networks

Figure 5.3 shows the organization of a concurrent computer. A number of processing nodes (N) communicate by means of an interconnection network. From Section 5.1 we know that the network must have a low latency to support fine-grain concurrent algorithms. We also know, from Section 5.2, that since VLSI technology is wire-limited, these networks are limited by the amount of wire required to construct them. In Section 5.3.1 I compare networks under the assumption of constant wire cost and show that low-dimension networks (e.g., a torus) offer lower latency than can be achieved with a high-dimensional interconnect (e.g., a binary n-cube). This surprising result strongly motivates the use of low-dimension k-ary n-cubes for the interconnection networks of concurrent computers.

---

[7] *Slow-clock logic* is a better name for this technique since it is the speed of the clocks relative to the circuit rather than their voltage level that results in an energy savings.

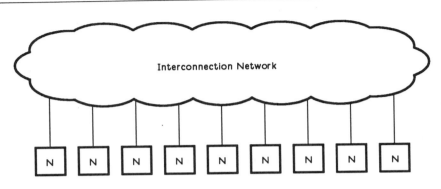

Figure 5.3: A Concurrent Computer

A deadlock-free routing algorithm for k-ary n-cube networks is required if these networks are to be useful. In Section 5.3.2 I develop a novel method for constructing deadlock-free routing algorithms and apply this method to several networks including k-ary n-cubes. To test these ideas, I have designed a VLSI chip that implements such a routing algorithm. The design and testing of this chip are described in Section 5.3.3.

5.3.1   Network Topology

Interconnection networks for concurrent computers have been studied intensely, and many different network topologies have been proposed. Tree networks have been proposed for use in concurrent computers [13]. However, it has been shown that most logical communication graphs do not map well onto a tree network topology[122]. A crossbar switch can be used to connect every node, P_i, to every other node, P_j. A crossbar has the desirable characteristic of being *non-blocking*. In a *non-blocking* network, any connection that describes a permutation of the processing nodes can be constructed without interference. Unfortunately crossbars are impractical for large systems because their wiring density grows as N^2. Benes [10] developed a non-blocking network for telephone systems that requires only $O(N \log N)$ switching elements. The Benes network has the disadvantage, however, that it requires a long time to configure for a particular permutation. In a concurrent computer where the pattern of communications varies dynamically, this long configuration time is unacceptable. Batcher's sorting network [7] is a more practical non-blocking network. While it requires $O(N \log^2 N)$ switching elements and has $O(\log^2 N)$ delay, it can be configured dynamically as messages are routed.

Most concurrent computers are constructed using blocking networks because the advantages of a non-blocking network are not sufficient to offset the $O(\log N)$ increased cost of a non-blocking network. The *Omega* network [82], a multiple stage shuffle-exchange network [124], is an example of such a blocking network. The *Omega* network has $O(N \log N)$ switching elements[8] and a delay of $O(\log N)$[9]. The *Omega* network is isomorphic to the *indirect* binary n-cube or *flip* network [8] [110]. The *direct* version of this network is the the binary n-cube [113], [98], [126]. The binary n-cube is a special case of the family of k-ary n-cubes, cubes with n dimensions and k nodes in each dimension.

Since most of the interconnection networks used for concurrent computers are isomorphic to binary n-cubes, a subset of k-ary n-cubes, in this section we restrict our attention to k-ary n-cube networks. It is the dimension of the network that is important, not the details of its topology. We refer to n as the *dimension* of the cube and k as the *radix*. Dimension, radix, and number of nodes are related by the equation

$$N = k^n, \quad \left(k = \sqrt[n]{N}, \quad n = \log_k N \right). \tag{5.12}$$

We can construct k-ary n-cubes with (approximately) the same number of nodes but with different dimensions. Figures 5.4-5.6 show three k-ary n-cube networks in order of decreasing dimension. Figure 5.4 shows a binary 6-cube (64 nodes). A 3-ary 4-cube (81 nodes) is shown in Figure 5.5. An 8-ary 2-cube (64 nodes), or torus, is shown in Figure 5.6. Each line in Figure 5.4 represents two communication channels, one in each direction, while each line in Figures 5.5 and 5.6 represents a single communication channel.

Networks have traditionally been analyzed under the assumption of constant channel bandwidth. Under this assumption each channel is one bit wide ($W = 1$) and has unit delay ($T_c = 1$). Thus, the constant bandwidth assumption favors networks with high dimensionality (e.g., binary n-cubes).

---

[8] Recall from Section 5.2 that it is the wiring density that is important, not the number of switching elements. I use the number of switching elements here for purposes of comparison only.

[9] The *Omega* network has $O(\log N)$ delay under the assumption that wire delay is independent of wire length. Again, I use this assumption here for purposes of comparison only. We have already seen that this assumption is not consistent with the characteristics of VLSI technology.

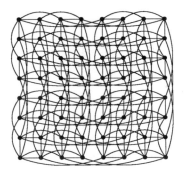

Figure 5.4: A Binary 6-Cube Embedded in the Plane

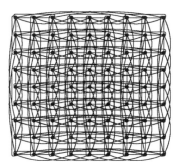

Figure 5.5: A Ternary 4-Cube Embedded in the Plane

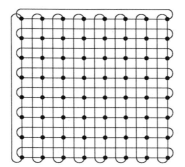

Figure 5.6: An 8-ary 2-Cube (Torus)

The constant bandwidth assumption, however, is not consistent with the properties of VLSI technology. Networks with many dimensions require more and longer wires than do low-dimensional networks. Thus, large dimensional networks cost more and run more slowly than low-dimensional networks. A realistic comparison of network topology must take both wire density and wire length into account.

In this section we compare the performance of k-ary n-cube interconnection networks using the following assumptions:

- Networks must be embedded into the plane[10].

- Nodes are placed systematically by embedding $\frac{n}{2}$ logical dimensions in each of the two physical dimensions. We assume that both n and k are even integers. The long end-around connections shown in Figure 5.6 can be avoided by folding the network as shown in Figure 5.22 on page 174.

- For networks with the same number of nodes, *wire density is held constant*. Each network is constructed with the same bisection width, B, the total number of wires crossing the midpoint of the network. To keep the bisection width constant, we vary the width, W, of the communication channels. We normalize to the bisection width of a bit-serial ($W = 1$) binary n-cube.

- The networks use *wormhole* routing, described in Section 5.3.2.

---

[10]If a three-dimensional packaging technology becomes available, the comparison changes only slightly.

- No more than a single bit is in transit on any wire at a given time.

- Channel delay, T_c, is a function of wire length, L. We begin by considering channel delay to be constant. Later, the comparison is performed for both logarithmic and linear wire delays; $T_c \propto \log L$ and $T_c \propto L$.

When k is even, the channels crossing the midpoint of the network are all in the highest dimension. For each of the $\sqrt{N}$ rows of the network, there are $k^{\left(\frac{n}{2}-1\right)}$ of these channels in each direction for a total of $2\sqrt{N}k^{\left(\frac{n}{2}-1\right)}$ channels. Thus, the bisection width, B, of a k-ary n-cube with W-bit wide communication channels is

$$B(k,n) = 2W\sqrt{N}k^{\left(\frac{n}{2}-1\right)}. \tag{5.13}$$

For a binary n-cube, $k = 2$, the bisection width is $B(2,n) = WN$. We set B equal to N to normalize to a binary n-cube with unit width channels, $W = 1$. The channel width, $W(k,n)$, of a k-ary n-cube with the same bisection width, B, follows from (5.13):

$$2W(k,n)\sqrt{N}k^{\left(\frac{n}{2}-1\right)} = N, \tag{5.14}$$

$$W(k,n) = \frac{\sqrt{N}}{2k^{\left(\frac{n}{2}-1\right)}} = \frac{\sqrt{N}}{2k^{\frac{n}{2}}k^{-1}} = \frac{k\sqrt{N}}{2\sqrt{N}} = \frac{k}{2}. \tag{5.15}$$

The peak wire density is greater than the bisection width in networks with $n > 2$ because the lower dimensions contribute to wire density. The maximum density, however, is bounded by

$$
\begin{aligned}
D_{\max} &= 2W\sqrt{N}\sum_{i=0}^{\frac{n}{2}-1} k^i \\
&= k\sqrt{N}\sum_{i=0}^{\frac{n}{2}-1} k^i \\
&= k\sqrt{N}\left(\frac{k^{\frac{n}{2}}-1}{k-1}\right) \\
&= k\sqrt{N}\left(\frac{\sqrt{N}-1}{k-1}\right) \\
&< \left(\frac{k}{k-1}\right)B.
\end{aligned}
\tag{5.16}
$$

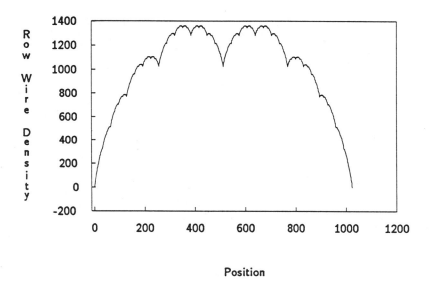

Figure 5.7: Wire Density vs. Position for One Row of a Binary 20-Cube

A plot of wire density as a function of position for one row of a binary 20-cube is shown in Figure 5.7. The density is very low at the edges of the cube and quite dense near the center. The peak density for the row is 1364 at position 341. Compare this density with the bisection width of the row, which is 1024. In contrast, a two-dimensional torus has a wire density of 1024 independent of position. One advantage of high-radix networks is that they have a very uniform wire density. They make full use of available area.

Each processing node has $2n$ channels each of which is $\frac{k}{2}$ bits wide. Thus, the number of pins per processing node is

$$N_p = nk. \tag{5.17}$$

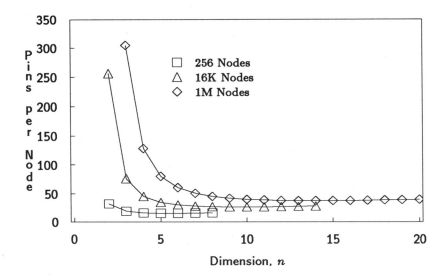

Figure 5.8: Pin Density vs. Dimension for 256, 16K, and 1M Nodes

A plot of pin density as a function of dimension for $N = 256$, 16K and 1M nodes[11] is shown in Figure 5.8. Low dimensional networks have the disadvantage of requiring many pins per processing node. A two-dimensional network with 1M nodes (not shown) requires 2048 pins and is clearly unrealizable. However, the number of pins decreases very rapidly as the dimension, n, increases. Even for 1M nodes, a dimension 4 node has only 128 pins. Recall from Section 5.2.1, however, that the wire density of the board under the chips becomes saturated before the maximum pin density of the chip is exceeded. Since all of the 1M node configurations have the same bisection width, $B = 1M$, these machines cannot be wired in a single plane.

---

[11]$1K = 1024$ and, $1M = 1K \times 1K = 1048576$.

Latency

From Section 5.1 we know that the latency of the network is the critical performance measure. Latency, T_l, is the sum of the latency due to the network and the latency due to the processing node,

$$T_l = T_{\text{net}} + T_{\text{node}}. \tag{5.18}$$

In this section we are concerned only with T_{net}. We will consider T_{node} in Section 5.4.

Network latency depends on the time required to drive the channel, T_c, the number of channels a message must traverse, D, and the number of cycles required to transmit the message across a single channel, $\frac{L}{W}$, where L is message length.

$$T_{\text{net}} = T_c \left(D + \frac{L}{W} \right) \tag{5.19}$$

If we select two processing nodes, P_i, P_j, at random, the average number of channels that must be traversed to send a message from P_i to P_j is given by the following three equations for the torus, the binary n-cube and general k-ary n-cubes:

$$D_t = \sqrt{N} - 1, \tag{5.20}$$

$$D_b = \frac{n}{2}, \tag{5.21}$$

$$D(k, n) = \left(\frac{k - 1}{2} \right) n. \tag{5.22}$$

The average latency of a k-ary n-cube is calculated by substituting (5.15) and (5.22), into (5.19)

$$T_{\text{net}} = T_c \left(\left(\frac{k - 1}{2} \right) n + \frac{2L}{k} \right). \tag{5.23}$$

Figure 5.9 shows the average network latency, T_{net}, as a function of dimension, n, for k-ary n-cubes with 2^8 (256), 2^{14} (16K), and 2^{20} (1M) nodes[12]. The left most data point in this figure corresponds to a torus ($n = 2$) and the right most data point corresponds to a binary n-cube ($k = 2$). This figure assumes constant wire delay, T_c, and a message length, L, of 150 bits. Although constant wire delay is unrealistic, this figure illustrates that even ignoring the dependence of wire delay on wire length, low-dimensional networks achieve lower latency than high-dimensional networks.

In general the lowest latency is achieved when the component of latency due to distance, D, and the component due to message length, $\frac{L}{W}$, are approximately equal, $D \approx \frac{L}{W}$. For the three cases shown in Figure 5.9, minimum latencies are achieved for $n = 2$, 4, and 5 respectively.

The length of the longest wire in the system, l_W, becomes a bottleneck that determines the rate at which each channel operates, T_c. The length of this wire is given by

$$l_W(k, n) = k^{\frac{n}{2}-1}. \tag{5.24}$$

If the wires are sufficiently short, delay depends logarithmically on wire length. If the channels are longer, they become limited by the speed of light, and delay depends linearly on channel length. Substituting (5.24) into (5.6) and (5.8) gives

$$T_c \propto \begin{cases} 1 + \log_e l_W = 1 + \left(\frac{n}{2} - 1\right)\log_e k & \text{logarithmic delay} \\[2mm] l_W = k^{\frac{n}{2}-1} & \text{linear delay.} \end{cases} \tag{5.25}$$

We substitute (5.25) into (5.23) to get the network latency for these two cases:

$$T_l \propto \begin{cases} \left(1 + \left(\frac{n}{2} - 1\right)\log_e k\right)\left(\left(\frac{k-1}{2}\right)n + \frac{2L}{k}\right) & \text{logarithmic delay} \\[2mm] \left(k^{\frac{n}{2}-1}\right)\left(\left(\frac{k-1}{2}\right)n + \frac{2L}{k}\right) & \text{linear delay.} \end{cases} \tag{5.26}$$

---

[12]For the sake of comparison we allow radix to take on non-integer values. For some of the dimensions considered, there is no integer radix, k, that gives the correct number of nodes. In fact, this limitation can be overcome by constructing a *mixed-radix cube*.

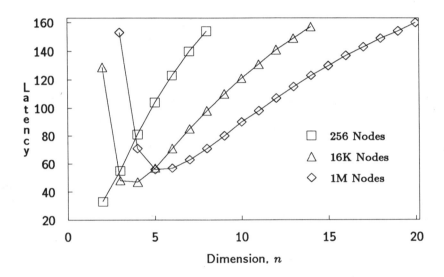

Figure 5.9: Latency vs. Dimension for 256, 16K, and 1M Nodes, Constant Delay

Figure 5.10 shows the average network latency as a function of dimension for
k-ary n-cubes with 2^8 (256), 2^{14} (16K), and 2^{20} (1M) nodes, assuming logarithmic wire delay and a message length, L, of 150. Figure 5.11 shows the same
data assuming linear wire delays. In both figures, the left most data point
corresponds to a torus $(n = 2)$ and the right most data point corresponds to a
binary n-cube $(k = 2)$.

In the linear delay case, Figure 5.11, a torus $(n = 2)$ always gives the lowest
latency. This is because a torus offers the highest bandwidth channels and the
most direct physical route between two processing nodes. Under the linear delay
assumption, latency is determined solely by bandwidth and by the physical
distance traversed. There is no advantage in having long channels.

Under the logarithmic delay assumption, Figure 5.10, a torus has the lowest latency for small networks $(N = 256)$. For the larger networks, the lowest latency
is achieved with slightly higher dimensions. With $N = 16K$, the lowest latency
occurs when n is three. With $N = 1M$, the lowest latency is achieved when n
is 5. It is interesting that assuming constant wire delay does not significantly
change this result. Recall that under the (unrealistic) constant wire delay assumption, Figure 5.9, the minimum latencies are achieved with dimensions of
2, 4, and 5 respectively.

The results shown in Figures 5.10 through 5.19 were derived by comparing networks under the assumption of constant wire cost to a binary n-cube with
$W = 1$. For small networks it is possible to construct binary n-cubes with
wider channels, and for large networks (e.g., $1M$ nodes) it may not be possible to construct a binary n-cube at all. In the case of small networks, the
comparison against binary n-cubes with wide channels can be performed by ex-
pressing message length in terms of the binary n-cube's channel width, in effect
decreasing the message length for purposes of comparison. The net result is the
same: lower-dimensional networks give lower latency. Even if we perform the
256 node comparison against a binary n-cube with $W = 16$, the torus gives the
lowest latency under the logarithmic delay model, and a dimension 3 network
gives minimum latency under the constant delay model. For large networks,
the available wire is less than assumed, so the effective message length should
be increased, making low dimensional networks look even more favorable.

In this comparison we have assumed that only a single bit of information is in
transit on each wire of the network at a given time. Under this assumption, the
delay between nodes, T_c, is equal to the period of each node, T_p. In a network
with long wires, however, it is possible to have several bits in transit at once. In
this case, the channel delay, T_c, is a function of wire length, while the channel
period, $T_p < T_c$, remains constant. Similarly, in a network with very short wires

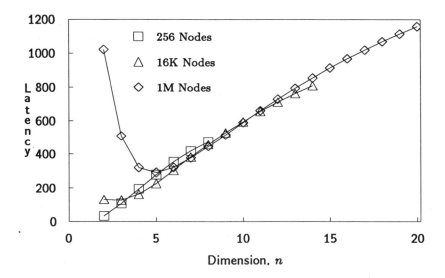

Figure 5.10: Latency vs. Dimension for 256, 16K, and 1M Nodes, Logarithmic Delay

we may allow a bit to ripple through several channels before sending the next bit. In this case, $T_p > T_c$. Separating the coefficients, T_c and T_p, (5.19) becomes

$$T_{\text{net}} = \left(T_c D + T_p \frac{L}{W}\right). \tag{5.27}$$

The net effect of allowing $T_c = T_p$ is the same as changing the length, L, by a factor of $\frac{T_p}{T_c}$ and does not change our results significantly.

When wire cost is considered, low-dimensional networks (e.g., tori) offer lower latency than high-dimensional networks (e.g., binary n-cubes). Intuitively, tori outperform binary n-cubes because they better match form to function. The logical and physical graphs of the torus are identical; Thus, messages always

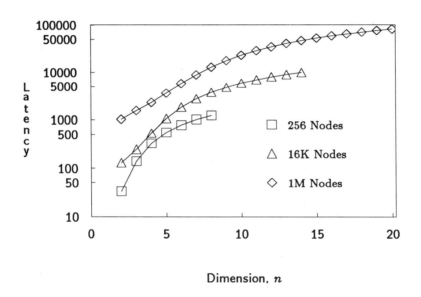

Figure 5.11: Latency vs. Dimension for 256, 16K, and 1M Nodes, Linear Delay

travel the minimum distance from source to destination. In a binary n-cube, on the other hand, the fit between form and function is not as good. A message in a binary n-cube embedded into the plane may have to traverse considerably more than the minimum distance between its source and destination.

Throughput

Throughput, another important metric of network performance, is defined as the total number of messages the network can handle per unit time. One method of estimating throughput is to calculate the capacity of a network, the total number of messages that can be in the network at once. Typically the maximum throughput of a network is some fraction of its capacity. The network capacity per node is the total bandwidth out of each node divided by the average number of channels traversed by each message. For k-ary n-cubes, the bandwidth out of each node is nW, and the average number of channels traversed is given by (5.22), so the network capacity per node is given by

$$
\Gamma(k,n) \; \propto \; \frac{nW(k,n)}{D(k,n)}
$$

$$
\propto \; \frac{n\left(\frac{k}{2}\right)}{\left(\frac{k-1}{2}\right)n} \tag{5.28}
$$

$$
\approx \; 1.
$$

The network capacity is independent of dimension. For a constant amount of wire, there is a constant network bandwidth.

Throughput will be less than capacity because contention causes some channels to block [27]. This contention also increases network latency. Let the traffic offered to the network by each node be $\lambda \frac{\text{bits}}{\text{cycle}}$. Consider a single dimension of the network as shown in Figure 5.12. The message rate on channels entering the dimension is $\lambda_E = \frac{\lambda}{L} \frac{\text{messages}}{\text{cycle}}$. The average message traverses $\frac{k-1}{2}$ channels in this dimension: one entering channel and $\sigma = \frac{k-3}{2}$ continuing channels. Thus, the rate on channels continuing in the dimensions is $\lambda_C = \sigma \lambda_E$. At the destination, each flit is serviced as soon as it arrives, so the service time at the sink is $T_{n-1} = \frac{L}{W} = \frac{2L}{k}$. Starting with T_{n-1} we will calculate the service time seen entering each preceding dimension.

Suppose the service time in dimension $i + 1$ is T_{i+1}. We wish to calculate the service time seen entering the previous dimension, T_i. The service time in the

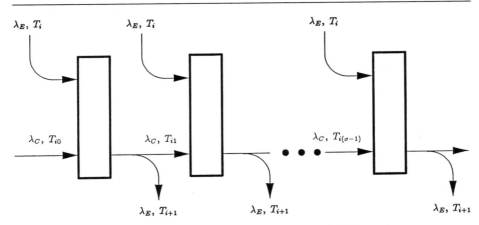

Figure 5.12: Contention Model for A Single Dimension

last continuing channel in this dimension is $T_{i(\sigma-1)} = T_{i+1}$. Once we know the service time for the jth channel, T_{ij}, the additional service time due to contention at the $j - 1^{\text{th}}$ channel is given by multiplying the probability of a collision, $\lambda_E T_{i0}$, by the expected waiting time for a collision, $\frac{T_{i0}}{2}$. Repeating this calculation σ times gives us T_{i0}.

$$T_{i(j-1)} = T_{ij} + \frac{\lambda_E T_{i0}^2}{2},$$

$$T_{i0} = T_{i+1} + \frac{\sigma \lambda_E T_{i0}^2}{2} = T_{i+1} + \frac{\lambda_C T_{i0}^2}{2}, \qquad (5.29)$$

$$= \frac{1 - \sqrt{1 - 2\lambda_C T_{i+1}}}{\lambda_C}.$$

Equation (5.29) is valid only when $\lambda_C < \frac{T_{i+1}}{2}$. If the message rate is higher than this limit, latency becomes infinite.

To calculate T_i we also need to consider the possibility of a collision on the entering channel.

$$T_i = T_{i0}\left(1 + \frac{\lambda_C T_{i0}}{2}\right). \qquad (5.30)$$

Parameter	256 Nodes			1024 Nodes		
Dimension	2	4	8	2	5	10
radix	16	4	2	32	4	2
Max Throughput	0.40	0.49	0.21	0.36	0.42	0.18
Latency $\lambda = 0.1$	43.9	121.	321.	45.3	128.	377.
Latency $\lambda = 0.2$	51.2	145.	648.	50.0	162.	NA
Latency $\lambda = 0.3$	64.3	180.	NA	59.0	221.	NA

Table 5.1: Maximum Throughput as a Fraction of Capacity and Blocking Latency in Cycles

If sufficient queueing is added to each network node, the service times do not increase, only the latency and (5.30) becomes.

$$T_i = \left(\frac{T_{i+1}}{1 - \frac{\lambda_C T_{n-1}}{2}} \right) \left(1 + \frac{\lambda_C T_{n-1}}{2} \right). \tag{5.31}$$

To be effective, the total queueing between the source and destination should be greater than the expected increase in latency due to blocking. One or two flits of queueing per stage is usually sufficient. The analysis here is pessimistic in that it assumes no queueing.

To find the maximum throughput of the network, the source service time, T_0, is set equal to the reciprocal of the message rate, λ_E, and equations (5.29) and (5.30) are solved for λ_E. The maximum throughput as a fraction of capacity for k-ary n-cubes with 256 and 1K nodes is tabulated in Table 5.1. Also shown is the total latency for L = 200bit messages at several message rates. The table shows that the additional latency due to blocking is significantly reduced as dimension is decreased.

Figure 5.13 compares measurements from a network simulator (points) to the latency predicted by (5.30) (lines). The simulation agrees with the prediction within a few percent until the network approaches saturation. When the network saturates, throughput levels off as shown in Figure 5.14. This plateau occurs because (1) the network is source queued, and (2) messages that encounter contention are blocked rather than aborted.

Intuitively, low-dimensional networks handle contention better because they use fewer channels of higher bandwidth and thus get better queueing performance. The shorter service times, $\frac{L}{W}$, of these networks results in both a lower

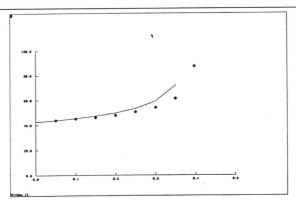

Figure 5.13: Latency vs. Traffic (λ) for 32-ary 2-cube, L=200bits. Solid line is predicted latency, points are measurements taken from a simulator.

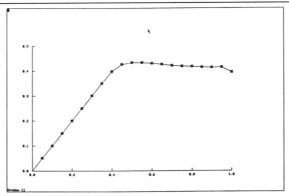

Figure 5.14: Actual Traffic vs. Attempted Traffic for 32-ary 2-cube, L=200bits.

probability of collision, and a lower expected waiting time in the event of a collision. Thus the blocking latency at each node is reduced quadratically as k is increased. Low-dimensional networks require more hops, $D = \frac{n(k-1)}{2}$, and have a higher rate on the continuing channels, λ_C. However, messages travel on the continuing channels more frequently than on the entering channels, thus most contention is with the lower rate channels. Having fewer channels of higher bandwidth also improves hot-spot throughput as described below.

Hot Spot Throughput

In many situations traffic is not uniform, but rather is concentrated into *hot spots*. A *hot spot* is a pair of nodes that accounts for a disproportionately large portion of the total network traffic. As described by Pfister [103] for a shared-memory computer, hot-spot traffic can degrade performance of the entire network by causing congestion.

The *hot-spot throughput* of a network is the maximum rate at which messages can be sent from one specific node, P_i, to another specific node, P_j. For a k-ary n-cube with deterministic routing, the hot-spot throughput, Θ_{HS}, is just the bandwidth of a single channel, W. Thus, under the assumption of constant wire cost we have

$$\Theta_{\text{HS}} = W = k - 1. \tag{5.32}$$

Low-dimensional networks have greater channel bandwidth and thus have greater hot-spot throughput than do high-dimensional networks. Intuitively, low-dimensional networks operate better under non-uniform loads because they do more resource sharing. In an interconnection network the resources are wires. In a high-dimensional network, wires are assigned to particular dimensions and cannot be shared between dimensions. For example, in a binary n-cube it is possible for a wire to be saturated while a physically adjacent wire assigned to a different dimension remains idle. In a torus all physically adjacent wires are combined into a single channel that is shared by all messages that must traverse the physical distance spanned by the channel.

5.3.2 Deadlock-Free Routing

Deadlock in the interconnection network of a concurrent computer occurs when no message can advance toward its destination because the queues of the message system are full [72]. Consider the example shown in Figure 5.15. The

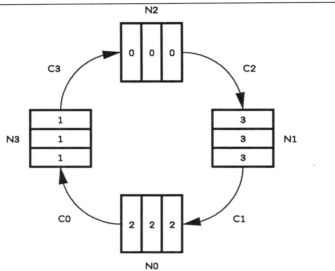

Figure 5.15: Deadlock in a 4-Cycle

queues of each node in the 4-cycle are filled with messages destined for the opposite node. No message can advance toward its destination; thus the cycle is deadlocked. In this locked state, no communication can occur over the deadlocked channels until exceptional action is taken to break the deadlock.

Definition 5.1 A flow control digit or *flit* is the smallest unit of information that a queue or channel can accept or refuse. Generally a *packet* consists of many *flits*. The unit of communication that is visible to the programmer is the *message*. A message may be composed of one or more *packets*, each of which carries its own routing and sequencing information in a header.

This complication of standard terminology has been adopted to distinguish between those flow control units that always include routing information - viz. packets - and those lower-level flow control units that do not - viz. flits. The literature on computer networks [127] has been able to avoid this distinction between packets and flits because most networks include routing information with every flow control unit; thus the flow control units are packets. That is not the case in the interconnection networks used by message-passing concurrent computers such as the Caltech Cosmic Cube [114].

The concurrent computer interconnection networks we are concerned with in this section are not store-and-forward networks. Instead of storing a packet completely in a node and then transmitting it to the next node, the networks we consider here use *wormhole routing*[13] [115]. With wormhole routing, only a few flits are buffered at each node. As soon as a node examines the header flit(s) of a packet, it selects the next channel on the route and begins forwarding flits down that channel. As flits are forwarded, the packet becomes spread out across the channels between the source and destination. It is possible for the first flit of a packet to arrive at the destination node before the last flit of the packet has left the source. Because most flits contain no routing information, the flits in a packet must remain in contiguous channels of the network and cannot be interleaved with the flits of other packets. When the header flit of a packet is blocked, all of the flits of a packet stop advancing and block the progress of any other packet requiring the channels they occupy. Because a single packet blocks many channels at once, preventing deadlock in a wormhole network is harder than preventing deadlock in a store-and-forward network.

I assume the following:

- Every packet arriving at its destination node is eventually consumed.

- A node can generate packets destined for any other node.

- The route taken by a packet is determined only by its destination and not by other traffic in the network.

- A node can generate packets of arbitrary length. Packets will generally be longer than a single flit.

- Once a queue accepts the first flit of a packet, it must accept the remainder of the packet before accepting any flits from another packet.

- An available queue may arbitrate between packets that request that queue space but may not choose amongst waiting packets.

- Nodes can produce packets at any rate subject to the constraint of available queue space (source queued).

The following definitions develop a notation for describing networks, routing functions, and configurations.

---

[13]A method similar to wormhole routing, called *virtual cut-through,* is described in [68]. Virtual cut-through differs from wormhole routing in that it buffers messages when they block, removing them from the network. With wormhole routing, blocked messages remain in the network.

Definition 5.2 An *interconnection network*, I, is a strongly connected *directed* graph, $I = G(N, C)$. The vertices of the graph, N, represent the set of processing nodes. The edges of the graph, C, represent the set of communication channels. Associated with each channel, c_i, is a queue with capacity $\text{cap}(c_i)$. The source node of channel c_i is denoted s_i and the destination node d_i.

Definition 5.3 A *routing function* $\mathbf{R} : C \times N \to C$ maps the current channel, c_c, and destination node, n_d, to the next channel, c_n, on the route from c_c to n_d, $\mathbf{R}(c_c, n_d) = c_n$. A channel is not allowed to route to itself, $c_c = c_n$. Note that this definition restricts the routing to be memoryless in the sense that a packet arriving on channel c_c destined for n_d has no memory of the route that brought it to c_c. However, this formulation of routing as a function from $C \times N$ to C has more memory than the conventional definition of routing as a function from $N \times N$ to C. Making routing dependent on the current channel rather than the current node allows us to develop the idea of channel dependence. Observe also that the definition of $\mathbf{R}$ precludes the route from being dependent on the presence or absence of other traffic in the network. $\mathbf{R}$ describes strictly deterministic and non-adaptive routing functions.

Definition 5.4 A *channel dependency graph*, D, for a given interconnection network, I, and routing function, $\mathbf{R}$, is a directed graph, $D = G(C, E)$. The vertices of D are the channels of I. The edges of D are the pairs of channels connected by $\mathbf{R}$:

$$E = \{(c_i, c_j) | \mathbf{R}(c_i, n) = c_j \text{ for some } n \in N\}. \tag{5.33}$$

Since channels are not allowed to route to themselves, there are no 1-cycles in D.

Definition 5.5 A *configuration* is an assignment of a subset of N to each queue. The number of flits in the queue for channel c_i will be denoted $\text{size}(c_i)$. If the queue for channel c_i contains a flit destined for node n_d, then $\text{member}(n_d, c_i)$ is true. A configuration is legal if

$$\forall c_i \in C, \ \text{size}(c_i) \leq \text{cap}(c_i). \tag{5.34}$$

Definition 5.6 A *deadlocked configuration* for a routing function, $\mathbf{R}$, is a nonempty legal configuration of channel queues such that

$$\forall c_i \in C, \ (\forall n \ni \text{member}(n, c_i), \ n = d_i \text{ and } c_j = \mathbf{R}(c_i, n) \Rightarrow \\ \text{size}(c_j) = \text{cap}(c_j)). \tag{5.35}$$

In this configuration no flit is one step from its destination, and no flit can advance because the queue for the next channel is full. A routing function, $\mathbf{R}$, is *deadlock-free* on an interconnection network, I, if no deadlock configuration exists for that function on that network.

Theorem 5.1 A routing function, $\mathbf{R}$, for an interconnection network, I, is deadlock-free iff there are no cycles in the channel dependency graph, D.

Proof:

$\Rightarrow$ Suppose a network has a cycle in D. Since there are no 1-cycles in D, this cycle must be of length two or more. Thus, one can construct a deadlocked configuration by filling the queues of each channel in the cycle with flits destined for a node two channels away, where the first channel of the route is along the cycle.

$\Leftarrow$ Suppose a network has no cycles in D. Since D is acyclic, one can assign a total order to the channels of C so that if $(c_i, c_j) \in E$ then $c_i > c_j$. Consider the least channel in this order with a full queue, c_l. Every channel, c_n, that c_l feeds is less than c_l, and thus does not have a full queue. Thus, no flit in the queue for c_l is blocked, and one does not have deadlock. ∎

Virtual Channels

Now that we have established this if-and-only-if relationship between deadlock and the cycles in the channel dependency graph, we can approach the problem of making a network deadlock-free by breaking the cycles. We can break such cycles by splitting each physical channel along a cycle into a group of *virtual channels*. Each group of virtual channels shares a physical communication channel; however, each virtual channel requires its own queue.

Consider for example the case of a unidirectional four-cycle as shown in Figure 5.16A, $N = \{n_0, \ldots, n_3\}$, $C = \{c_0, \ldots, c_3\}$. The interconnection graph, I, is shown on the left and the dependency graph, D, is shown on the right. We pick channel c_0 to be the dividing channel of the cycle and split each channel into high virtual channels, $c_{10}, \ldots, c_{13}$, and low virtual channels, $c_{00}, \ldots, c_{03}$, as shown in Figure 5.16B.

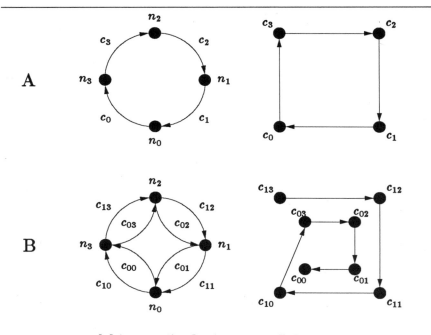

I: Interconnection Graph   D: Dependency Graph

Figure 5.16: Breaking Deadlock with Virtual Channels

When a packet enters the network it is routed on the high channels until it passes through node zero. After passing through node zero, packets are routed on the low channels. Channel c_{00} is not used. We now have a total ordering of the virtual channels according to their subscripts: $c_{13} > c_{12} > c_{11} > c_{10} > c_{03} > c_{02} > c_{01}$. Thus, there is no cycle in D, and the routing function is deadlock-free.

Many deadlock-free routing algorithms have been developed for store-and-forward computer communications networks [50], [51], [60], [91], [132], [133]. These algorithms are all based on the concept of a *structured buffer pool*. The packet buffers in each node of the network are partitioned into classes, and the assignment of buffers to packets is restricted to define a partial order on buffer classes. The structured buffer pool method has in common with the virtual channel method that both prevent deadlock by assigning a partial order to resources. The two methods differ in that the structured buffer pool approach restricts the assignment of buffers to packets, while the virtual channel approach restricts the routing of messages. Either method can be applied to store-and-forward networks, but the structured buffer pool approach is not directly applicable to wormhole networks, since the flits of a packet cannot be interleaved.

In the next section, virtual channels are used to construct a deadlock-free routing algorithm for k-ary n-cubes. In [24] algorithms are developed for cube-connected cycles and shuffle-exchange networks as well.

k-ary n-cubes

The E-cube routing algorithm [81],[126] guarantees deadlock-free routing in binary n-cubes. In a cube of dimension d, we denote a node as n_k where k is an d-digit binary number. Node n_k has d output channels, one for each dimension, labeled $c_{0k}, \ldots, c_{(d-1)k}$. The E-cube algorithm routes in decreasing order of dimension. A message arriving at node n_k destined for node n_l is routed on channel c_{ik}, where i is the position of the most significant bit in which k and l differ. Since messages are routed in order of decreasing dimension and hence decreasing channel subscript, there are no cycles in the channel dependency graph, and E-cube routing is deadlock-free.

Using the technique of virtual channels, this routing algorithm can be extended to handle all k-ary n-cubes. Rings and toroidal meshes are included in this class of networks. This algorithm can also handle mixed radix cubes. Each node of a k-ary n-cube is identified by an n-digit radix k number. The ith digit of the number represents the node's position in the ith dimension. For example, the center node in the 3-ary 2-cube of Figure 5.17 is n_{11}. Each channel is identified

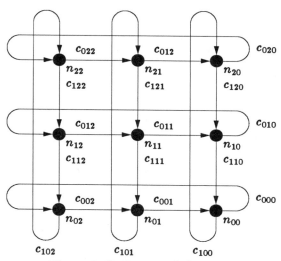

Figure 5.17: 3-ary 2-Cube

by the number of its source node and its dimension. For example, the dimension 0 (horizontal) channel from n_{11} to n_{10} is c_{011}. To break cycles, we divide each channel into an upper and lower virtual channel. The upper virtual channel of c_{011} will be labeled c_{0111}, and the lower virtual channel will be labeled c_{0011}. Internal channels are labeled with a dimension higher than the dimension of the cube. To assure that the routing is deadlock-free, we restrict it to route through channels in order of descending subscripts.

As in the E-cube algorithm, we route in order of dimension, most significant dimension first. In each dimension, i, a message is routed in that dimension until it reaches a node whose subscript matches the destination address in the ith position. The message is routed on the high channel if the ith digit of the destination address is greater than the ith digit of the present node's address. Otherwise the message is routed on the low channel. It is easy to see that this routing algorithm routes in order of descending subscripts and is thus deadlock-free.

Formally, we define the routing function:

$$\mathbf{R_{KNC}}(c_{dvn}, n_j) = \begin{cases} c_{d1(n-r^d)} \text{ if } (\text{dig}(n,d) < \text{dig}(j,d)) \wedge (\text{dig}(n,d) = 0), \\ c_{d0(n-r^d)} \text{ if } (\text{dig}(n,d) > \text{dig}(j,d)) \vee (\text{dig}(n,d) = 0), \\ c_{i1(n-r^d)} \text{ if } (\forall k > i, \text{dig}(n,k) = \text{dig}(j,k)) \wedge \\ \qquad\qquad (\text{dig}(n,i) = \text{dig}(j,i)), \end{cases} \quad (5.36)$$

where $\text{dig}(n,d)$ extracts the dth digit of n, and r is the radix of the cube. The subtraction, $n - r^d$, is performed so that only the dth digit of the address n is affected.

Assertion **5.1** The routing function, $\mathbf{R_{KNC}}$, correctly routes messages from any node to any other node in a k-ary n-cube.

Proof: By induction on dimension, d.

For $d = 1$, a message, destined for n_j, enters the system at n_i on the internal channel, c_{d0i}. If $i < j$, the message is forwarded on channels, $c_{01i}, \ldots, c_{010}, c_{00r}, \ldots, c_{00(j+1)}$ to node n_j. If $i > j$, the path taken is, $c_{00i}, \ldots, c_{00(j+1)}$. In both cases the route reaches node n_j.

Assume that the routing works for dimensions $\leq d$ Then for dimension $d + 1$ there are two cases. If $\text{dig}(i,d) = \text{dig}(j,d)$, then the message is routed around the most significant cycle to a node $n_k \ni \text{dig}(k,d) = \text{dig}(j,d)$, as in the $d = 1$ case above. If $\text{dig}(i,d) = \text{dig}(j,d)$, then the routing need be performed only in dimensions d and lower. In each of these cases, once the message reaches a node, $n_k, \ni \text{dig}(k,d) = \text{dig}(j,d)$, the third routing rule is used to route the message to a lower-dimensional channel. The problem has then been reduced to one of dimension $\leq n$, and the routing reaches the correct node by induction. ∎

Assertion **5.2** The routing function $\mathbf{R_{KNC}}$ on a k-ary n-cube interconnection network, I, is deadlock-free.

Proof: Since routing is performed in decreasing order of channel subscripts, $\forall c_i, c_j, n_c \ni \mathbf{R}(c_i, n_c) = c_j \Rightarrow i > j$, the channel dependency graph, D, is acyclic. Thus by Theorem 5.1 the route is deadlock-free. ∎

Figure 5.18: Photograph of the Torus Routing Chip

Figure 5.19: A Packaged Torus Routing Chip

5.3.3 The Torus Routing Chip

I have developed the torus routing chip (TRC) as a demonstration of the use of virtual channels for deadlock-free routing. Shown in Figures 5.18 and 5.19, the TRC is a $\approx 10,000$-transistor chip implemented in 3μ CMOS technology and packaged in an 84-lead pin-grid array. It provides deadlock-free packet communications in k-ary n-cube (torus) networks with up to $k = 256$ nodes in each dimension. While primarily intended for $n = 2$-dimensional networks, the chips can be cascaded to handle arbitrary n-dimensional networks using $\frac{n}{2}$ TRCs at each processing node. TRCs have been fabricated and tested.

Even if only two dimensions are used, the TRC can be used to construct concurrent computers with up to 2^{16} nodes. It would be very difficult to distribute a global clock over an array of this size [42]. To avoid this problem, the TRC is entirely self-timed [111], thus permitting each processing node to operate at its own rate with no need for global synchronization. Synchronization, when required, is performed by arbiters in the TRC.

To reduce the latency of communications that traverse more than one channel, the TRC uses *wormhole* [115] routing rather than *store-and-forward* routing. Instead of reading an entire packet into a processing node before starting trans-

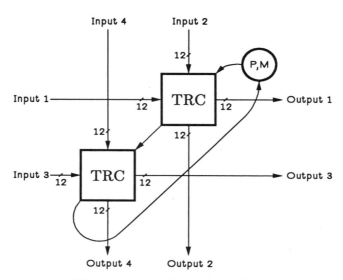

Figure 5.20: A Dimension 4 Node

mission to the next node, the TRC forwards each byte of the packet to the next node as soon as it arrives. Wormhole routing thus results in a message latency that is the *sum* of two terms, one of which depends on the message length, L, and the other of which depends on the number of communication channels traversed, D. Store-and-forward routing gives a latency that depends on the product of L and D. Another advantage of wormhole routing is that communications do not use up the memory bandwidth of intermediate nodes. A packet does not interact with the processor or memory of intermediate nodes along its route. Packets remain strictly within the TRC network until they reach their destination.

System Design

The torus routing chip (TRC) can be used to construct arbitrary k-ary n-cube interconnection networks. Each TRC routes packets in two dimensions, and the chips are cascadable as shown in Figure 5.20 to construct networks of dimension greater than two. The first TRC in each node routes packets in the first two dimensions and strips off their address bytes before passing them to the second TRC. This next chip then treats the next two bytes as addresses in the next

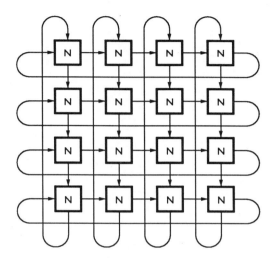

Figure 5.21: A Torus System

two dimensions and routes packets accordingly. The network can be extended to any number of dimensions.

A block diagram of a 2-dimensional message-passing concurrent computer constructed around the TRC is shown in Figure 5.21. Each node consists of a processor, its local memory, and a TRC. Each TRC in the torus is connected to its processor by a processor input channel and a processor output channel. Connections on the edges of the torus wrap around to the opposite edge. One can avoid the long end-around connection by folding the torus, as shown in Figure 5.22.

A *flit* in the TRC is a byte whose 8 bits are transmitted in parallel. The X and Y channels each consist of 8 data lines and 4 control lines. The 4 control lines are used for separate request/acknowledge signal pairs for each of two virtual channels. The processor channels are also 8 bits wide, but have only two control lines each.

The packet format is shown in Figure 5.23. A packet begins with two address bytes. The bytes contain the relative X and Y addresses of the destination node. The relative address in a given direction, say X, is a count of the number

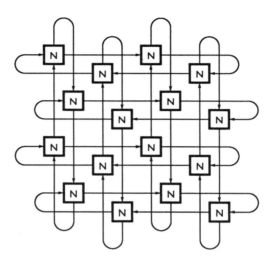

Figure 5.22: A Folded Torus System

of channels that must be traversed in the X direction to reach a node with the same X address as the destination. After the address comes the data field of the packet. This field may contain any number of *non-zero* data bytes. The packet is terminated by a zero tail byte. Later versions of the TRC may use an extra bit to tag the tail of a packet, and might also include error checking.

The TRC network routes packets first in the X direction, then in the Y direction. Packets are routed in the direction of decreasing address, decrementing the relative address at each step. When the relative X address is decremented to zero, the packet has reached the correct X coordinate. The X address is then stripped from the packet, and routing is initiated in the Y dimension. When the Y address is decremented to zero, the packet has reached the destination node. The Y address is then stripped from the packet, and the data and tail bytes are delivered to the node.

Each of the X and Y physical channels is multiplexed into two virtual channels. In each dimension packets begin on virtual channel 1. A packet remains on virtual channel 1 until it reaches its destination or address zero in the direction of routing. After a packet crosses address zero, it is routed on virtual channel 0. The address 0 origin of the torus network in X and Y is determined by two

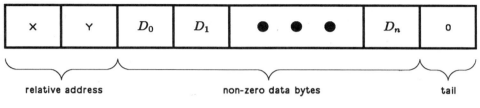

Figure 5.23: Packet Format

input pins on the TRC. The effect of this routing algorithm is to break the channel dependency cycle in each dimension into a two-turn spiral similar to that shown in Figure 5.16 on page 166. Packets enter the spiral on the outside turn and reach the inside turn only after passing through address zero.

Each virtual channel in the TRC uses the 2-cycle signaling convention shown in Figure 5.24. Each virtual channel has its own request (R) and acknowledge (A) lines. When $R = A$, the receiver is ready for the next flit (byte). To transfer information, the sender waits for $R = A$, takes control of the data lines, places data on the data lines, toggles the R line, and releases the data lines. The receiver samples data on each transition of R line. When the receiver is ready for the next byte, it toggles the A line.

The protocol allows both virtual channels to have requests pending. The sending end does not wait for any action from the receiver before releasing the channel. Thus, the other virtual channel will never wait longer than the data transmission time to gain access to the channel. Since a virtual channel always releases the physical channel after transmitting each byte, the arbitration is fair. If both channels are always ready, they will alternate bytes on the physical channel.

Consider the example shown in Figure 5.25. Virtual channel X1 gains control of the physical channel, transmits one byte of information, and releases the channel. Before this information is acknowledged, channel X0 takes control of the channel and transmits two bytes of information. Then X1, having by then been acknowledged, takes the channel again.

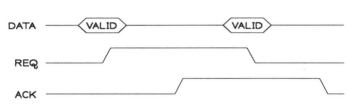

Figure 5.24: Virtual Channel Protocol

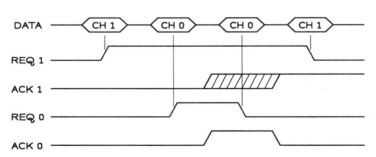

Figure 5.25: Channel Protocol Example

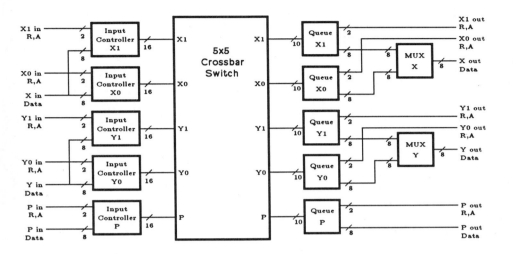

Figure 5.26: TRC Block Diagram

Logic Design

As shown in Figure 5.26, the TRC consists of five input controllers, a five by five crossbar switch, five output queues, and two output multiplexers. There is one input controller and one output controller for each virtual channel. The output multiplexers serve to multiplex two virtual channels onto a single physical channel.

The input controller is responsible for packet routing. When a packet header arrives, the input controller selects the output channel, adjusts the header by decrementing and sometimes stripping the byte, and then passes all bytes to the crossbar switch until the tail byte is detected.

The input controller, shown in Figure 5.27, consists of a datapath and a self-timed state machine. The datapath contains a latch, a zero checker, and a decrementer. A state latch, logic array, and control logic comprise the state machine. When the request line for the channel is toggled, data are latched, and the zero checker is enabled. When the zero checker makes a decision, the logic array is enabled to determine the next state, the selected crossbar channel,

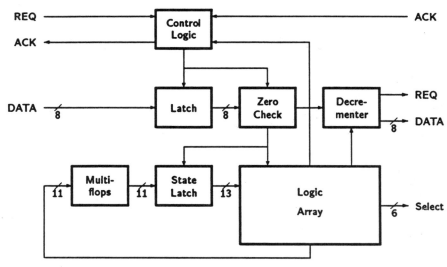

Figure 5.27: Input Controller Block Diagram

and whether to strip, decrement, or pass the current byte. When the required operation has been completed, possibly requiring a round trip through the crossbar, the state and selected channel are saved in cross-coupled multi-flops and the logic array is precharged.

The input controller and all other internal logic operate using a 4-cycle signaling convention [111]. One function of the state machine control logic is to convert the external 2-cycle convention into the on-chip 4-cycle convention. The signals are converted back to 2-cycle at the output pads.

The crossbar switch performs the switching and arbitration required to connect the five input controllers to the five output queues. A single crosspoint of the switch is shown in Figure 5.28. A two-input interlock (mutual-exclusion) element in each crosspoint arbitrates requests from the current input channel (row) with requests from all lower channels (rows). The interlock elements are connected in a priority chain so that an input channel must win the arbitration in the current row and all higher rows before gaining access to the output channel (column).

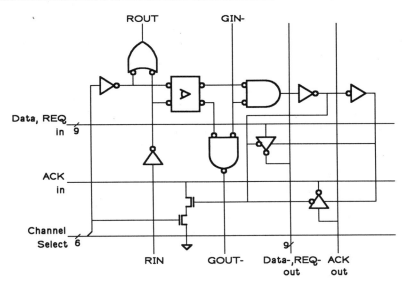

Figure 5.28: Crosspoint of the Crossbar Switch

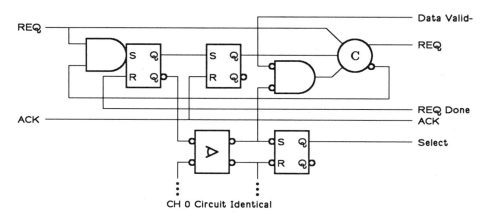

Figure 5.29: Output Multiplexer Control

The output queues buffer data from the crossbar switch for output. The queues are of length four. While shorter queues would suffice to decouple input and output timing, the longer queues also serve to smooth out the variation in delays due to channel conflicts.

Each output multiplexer performs arbitration and switching for the virtual channels that share a common physical channel. As shown in Figure 5.29, a small self-timed state machine sequences the events of placing the data on the output pads, asserting request, and removing the output data. An interlock element is used to resolve conflicts between channels for the data pads.

To interface the on-chip equipotential region to the off-chip equipotential region that connects adjacent chips, self-timed output pads (Figure 7.22 in [111]) are used. A Schmidt Trigger and exclusive-OR gate in each of these pads signals the state machine when the pad is finished driving the output. These completion signals are used to assure that the data pads are valid before the request is asserted and that the request is valid before the data are removed from the pads and the channel released.

Experimental Results

The design of the TRC began in August 1985. The chip was completely designed and simulated at the transistor level before any layout was performed.

The circuit design was described using CNTK, a language embedded in C [26], and was simulated using MOSSIM [14]. A subtle error in the self-timed controllers was discovered at the circuit level before any time-consuming layout was performed. Once the circuit design was verified, the TRC was laid out in the new MOSIS scalable CMOS technology [134] using the Magic system [96]. A second circuit description was generated from the artwork and six layout errors were discovered by simulation of the extracted circuit. The verified layout was submitted to MOSIS for fabrication in September 1985.

The first batch of chips was completed the first week of December but failed to function because of fabrication errors. A second run of chips (same design), returned the second week of December, contained some fully functional chips.

Performance measurements on the chips are shown in Figure 5.30. To measure the maximum channel rate, output request and acknowledge lines were tied together, and input acknowledge was inverted and fed back into input request. In this configuration the chip runs at a maximum speed, shown in Figure 5.30A, of 4MHz. The delays from input request to output request and input acknowledge, shown in Figure 5.30B, are 150ns and 250ns respectively. Data propagation time from input to output (not shown) was measured to be 60ns for both rising and falling edges. Thus data are set up 90ns ahead of the output request. Data hold time, shown in Figure 5.30C, is 20ns.

Tau model calculations suggest that a redesigned TRC should operate at 20MHz and have an input to output delay of 50ns. The redesign would involve decoupling the timing of the input controller by placing single-stage queues between the input pads and input controller and between the input controller and the crossbar switch. The input controller would be modified to speed up critical paths.

Summary

Communication between nodes of a concurrent computer need not be slower than the communication between the processor and memory of a conventional sequential computer. By using byte-wide datapaths and wormhole routing, the TRC provides node-to-node communication times that approach main memory access times of sequential computers. Communications across the diameter of a network, however, may require substantially longer than a memory access time.

The TRC serves as still another counterexample to the myth that self-timed systems are more complex than synchronous systems. The design of the TRC is not significantly more complex than a synchronous design that performs the

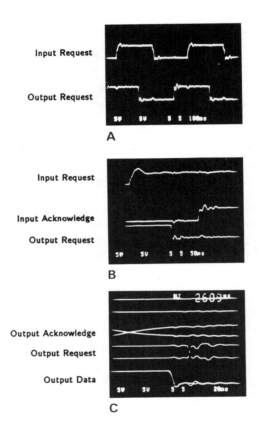

Figure 5.30: TRC Performance Measurements

same function. As for speed, the TRC is probably faster than a synchronous chip, since each chip can operate at its full speed with no danger of timing errors. A synchronous chip is generally operated at a slower speed that reflects the timing of a worst-case chip and adds a timing margin.

5.4   A Message-Driven Processor

In Section 5.3 we investigated means of minimizing message latency, T_l, by choosing the proper dimension interconnection network and proper routing strategy. We ignored, however, the contribution of the processing node to latency: T_{node} in (5.18). In this section I present novel architectural features that minimize T_{node} by matching the behavior of the processor to the object-based model of computation described by function (2.1).

In a concurrent computer built around a conventional instruction processor, interpreting a message is a time-consuming process. First, the processor responds to an interrupt informing it that a message has arrived. Next, the message is fetched from memory, and the method to be executed in response to the message is determined. Finally, after executing ≈ 100 instructions, the processor begins execution of the method. If the execution of the method involves sending a message, another cumbersome instruction sequence is required to initiate the send. The latency introduced by performing these message receives and sends in software is intolerable in a system where the average method is only 10 instructions long.

Instead of nesting the instruction fetch-decode-execute loop of a conventional processor inside the receive-dispatch-execute loop required to process a message, a message driven processor directly interprets messages. A level of interpretation is removed; messages are the instructions of a message-driven processor [28].

When a message arrives at a processing node, the processor performs the following steps:

Reception: Upon message arrival the message is immediately removed from the network. The message is buffered if the processor is busy and is received when the processor becomes idle. Reception and buffering of messages are performed by hardware. The current message is placed in a receive register to allow the processor fast access to arguments.

Method Lookup: Once a message has been copied into the receive register, the method corresponding to the message is determined by examining the message selector and the class of the receiver. An instruction translation lookaside buffer (ITLB) [22] is used to speed the translation of messages into methods.

Execution: Methods are either primitive or defined. Primitive methods, small integer add for example, are performed directly by the processor. They generally involve modifying the contents of an object and/or transmitting a reply message.

Defined methods create a context and specify a sequence of actions. Actions are similar to subroutines on a conventional processor. They are executed by sending a sequence of messages. Some of the sends performed during the execution of a defined method are handled locally. They are simply instructions. Sends to objects outside the current processing node result in sending a message over the network. Addressing modes are provided to allow fast access to the fields of the current message, acquaintances of the receiver, and the contents of the context during the execution of an action.

If a method consists of more than a single action, the context is retained, and the messages transmitted by the method are directed to reply to the context. A pointer to the next sequence of messages to be executed for the method is stored in the context. After the final action of a method, the processor sends a reply to the object specified in the Reply To field of the original message unless this field is nil.

The classes (data types) and the operations supported by a processing node may vary amongst nodes. As described in Section 5.5, some processing nodes may be object experts specialized to store and operate on a particular class of objects.

5.4.1 Message Reception

The format of a message is shown in Figure 5.31. Each message contains the following fields:

Receiver: The identifier of the object to which the message is directed.

Selector: The name of the message. The selector, together with the class of the receiver, determines what method is to be executed in response to the

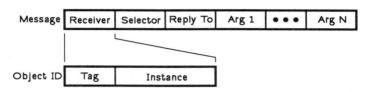

Figure 5.31: Message Format

message. If the message has a nil receiver, the selector directly determines the method[14].

Reply to: The object that is to receive the reply from this message. If this field is nil, no reply is expected.

Arguments: Object identifiers for the arguments of the message, if any.

As shown in the lower portion of Figure 5.31, each object identifier consists of two fields. The Tag field specifies whether the object is a primitive or a reference object and, if the object is a primitive, specifies its class. If the object is a primitive, the Instance field is the object itself. For example, if the Tag field specifies that the object is of class Small Integer, the Instance field contains the integer. For reference objects, the instance field contains a pointer to the object in object space. The object pointer is translated into a node number by the global mail system and into an address within the node by the local mail system. The class of a reference object is found within the object itself.

The process of message reception is illustrated in Figure 5.32. If the processor is idle when a message arrives from the network, the message is read directly into a receive register. The receive register contains slots for the receiver, selector, and reply to fields of the message, as well as four arguments. Additional arguments are stored in memory at a location referenced by the argument pointer.

If the processor is busy when a message arrives, the message is automatically buffered in memory. Message buffer memory access takes priority over processor memory access, since it is critical to network performance that a message be removed from the network as soon as it arrives at its destination node. Dedicated registers point to the head and tail of a message queue in memory. When

---

[14]Messages from the network will never have a nil receiver. Messages executed as the instructions of a defined method, however, may have a nil receiver.

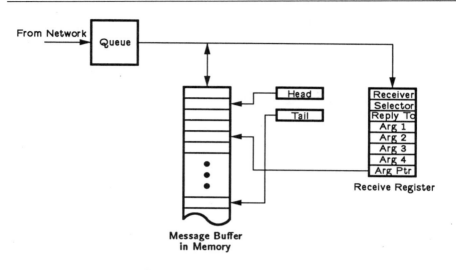

Figure 5.32: Message Reception

the processor becomes idle, the message at the head of the queue is removed from the queue and copied into the receive register.

The use of special purpose hardware to remove messages from the network, buffer them in memory, and load messages into a processor register has two significant performance advantages. Since messages are quickly removed from the network, network performance is improved; if messages were left for any period of time with their tails blocking network channels, severe network congestion could result. Also, message latency is reduced since the time for a conventional processor to respond to a network interrupt and load the message is eliminated. If the processor is idle, the message is loaded as soon as it arrives.

5.4.2   Method Lookup

Once a message is received, the first step in interpreting the message is to look up the method specified by the selector and the class of the receiver. If the receiver is a primitive, its class is encoded in the tag part of the object ID and is already in the receive register. If the receiver is a reference object, the class of the object must be fetched. Part of the object's class is a table of selectors understood by that class and the method corresponding to each selector. This

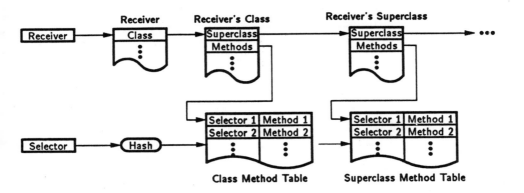

Figure 5.33: Method Lookup

table is searched, by hashing, to find the selector in the received message. If the selector is found, the corresponding method is executed. Otherwise, the object superclass is checked, then the superclass' superclass, and so on, as shown in Figure 5.33.

Method lookup can be accelerated by using an ITLB as shown in Figure 5.34 [22]. The ITLB is an associative memory that associates selector and class with the corresponding method. Each entry in the ITLB corresponds to a unique method and contains three fields:

Key: The selector and class that specify the method.

Primitive Bit: Specifies whether the method is primitive or defined.

Method: How the method is to be accomplished. For a primitive method, this field determines which primitive operation the processor is to perform. For a defined method, this field contains the object ID of the method.

Method lookup using the ITLB proceeds in three steps. First, the class of the receiver is obtained and concatenated with the selector to form a key into the ITLB. The ITLB, an associative memory, attempts to find an entry matching this key. If an ITLB entry is found, then the method field and primitive bit are read from the ITLB. Otherwise, a conventional method lookup must be

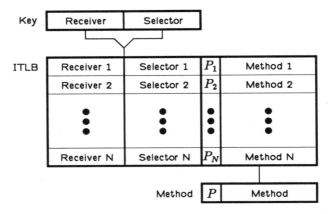

Figure 5.34: Instruction Translation Lookaside Buffer

performed as described above. All primitive instructions are permanently stored in the ITLB.

The memory requirements for a message-driven processor are quite modest. A processing element need not keep a complete class description for each class of objects it contains. In fact, a processing element need not keep any code resident at all. When a method is referenced, it can be copied over the network. When an ITLB miss occurs, method lookup can be spread across a number of processing elements[15]. Performance can be enhanced, however, by maintaining redundant copies of some methods and class descriptions. For example, it would be beneficial to maintain local copies of each method referenced in the ITLB.

5.4.3   Execution

A context object, shown in Figure 5.35, is created at the start of a defined method and controls the execution of the method. The receiver, reply-to object, and arguments from the message as well as the method pointer from the method lookup operation are copied into the context. The context contains the instruction pointer (IP) that sequences the instructions in the method. Local

---

[15]It is important that only primitive or guaranteed resident methods are used for method lookup. Otherwise, the system could get into an infinite loop of looking up a method required to look up a method etc....

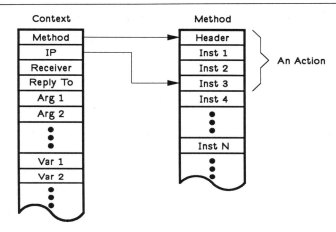

Figure 5.35: A Context

variables are also held in the context. A context cache as described in [22] can be used to provide fast allocation of and access to contexts.

To execute a method, the processor fetches the instruction referenced by the IP, executes the instruction, and increments the IP to proceed to the next instruction. Each instruction conceptually sends a message, by specifying the receiver, selector, and arguments. Many of the instructions, however, will not result in actually sending a message, but instead will modify the contents of a local object or alter execution of the method by modifying the IP. The processor executes these instructions directly.

Since all instructions are message sends, a processor that interprets a single instruction would suffice. However, to improve the efficiency of commonly used primitives and control instructions, certain messages will be more compactly encoded. A set of possible instructions is as follows:

send: The SEND instruction sends a message to the specified receiver. Addressing modes are provided to represent compactly the receiver, selector, and arguments. If the receiver is nil, a constant, or a local object, the operation will be performed directly and no send will occur. Otherwise, the fields of the message will be assembled and the message transmitted over the network.

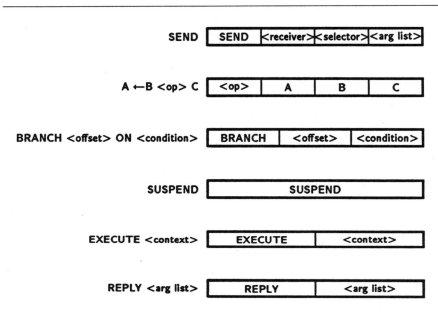

Figure 5.36: Instruction Formats

control instructions: These instructions can be thought of as sending messages to the current context. BRANCH conditionally alters the value of the IP. SUSPEND halts execution but preserves the context so that a reply can resume execution. EXECUTE suspends the current context and begins execution of another context. REPLY sends a reply message with a specified argument list to the object specified by the Reply-To field of the context and then deletes the current context.

common messages: Commonly used messages such as MOVE, at:put:, or + may be encoded directly to save space. Often these instructions combine several conceptual sends into a single instruction. For example, the instruction A ←B + C sends the message + C to B and then stores the result in A, conceptually sending an at:put: message to the object containing A.

Possible formats for these instructions are shown in Figure 5.36. Each instruction consists of an opcode field and zero or more operand fields. Each operand field contains an operand specifier that describes the operand using one of four addressing modes.

converts cycles in the channel dependency graph, D, into spirals, thus avoiding deadlock.

I have developed the Torus Routing Chip (TRC), described in Section 5.3.3, to demonstrate the feasibility of the type of network described in this chapter. The TRC combines many novel features.

- It is completely self-timed [111].

- It uses wormhole routing [115].

- It implements the virtual channel deadlock-free routing algorithm [24] in hardware.

TRCs have been fabricated and they operate properly.

In addition to minimizing network latency, the latency of each processing node must also be minimized by matching the architecture of the processor to the semantics of the programming model. Section 5.4 outlines the architecture of a message-driven processing element that responds directly to messages rather than interpreting messages using a conventional instruction processor.

To take advantage of the performance offered by specialization while at the same time retaining flexibility, processing elements can be specialized to operate on a single class of object. These *object experts*, Section 5.5, by accelerating common object classes, improve the performance of all applications using those classes. Object experts also promote locality by storing the objects local to the hardware that modifies them.

Chapter 6

Conclusion

The performance of computers can be made incrementally extensible by exploiting VLSI technology to build concurrent computers, ensembles of processing nodes connected by a network. These concurrent computers can be programmed by combining concurrent data structures. The problems of communication and synchronization are encapsulated in the data structure, leaving the programmer free to concentrate on problems specific to his/her application.

This thesis has developed a paradigm for programming concurrent computers: concurrent data structures. To describe concurrent data structures, a programming notation, Concurrent Smalltalk (CST), has been developed incorporating the concept of a distributed object. A distributed object is a single object consisting of a collection of constituent objects, each of which can receive messages sent to the distributed object. Thus distributed objects can process many messages simultaneously. They are the foundation upon which concurrent data structures are built.

The balanced cube is a concurrent data structure for ordered sets. It achieves concurrency by eliminating the root bottleneck of tree-based data structures. A balanced cube has no root; all nodes are equals. An ordered set is represented in a balanced cube by mapping elements of the set to right subcubes of the balanced cube using a Gray code. The VW search algorithm, based on the distance properties of the Gray code, searches a balanced cube in logarithmic time. This search algorithm can be initiated from any node and will uniformly distribute activity over the nodes of the cube. The B-cube is an extension of the balanced cube that stores several data in each node to match the grain size of the data structure to the grain size of a particular computer. The balanced cube is an example of a concurrent data structure that differs markedly from its sequential counterparts.

Concurrent graph data structures can be used to solve many combinatorial problems. In Chapter 4 concurrent algorithms for the shortest path problem, the max-flow problem, and graph partitioning were developed. These graph algorithms illustrate many of the synchronization problems encountered in concurrent programming. Consider, for example, the shortest path problem. Dijkstra's sequential algorithm [29] cannot directly be made concurrent because it depends on a total order of events. It is too tightly synchronized. A concurrent algorithm due to Chandy and Misra [15] that relaxes this ordering of events but introduces no other synchronization may require exponential time because it is too loosely synchronized.

The concurrent algorithms described here are characterized by short messages and short methods. Supporting this fine-grain concurrency requires a low-latency interconnection network for efficient execution. Because VLSI technology is wire-limited, alternative architectures must be compared keeping wire cost constant. Consider the family of k-ary n-cube networks: networks with n dimensions and k processors in each dimension. High-dimensional networks with narrow channels are compared against low-dimensional networks with wide channels. The minimum latency occurs when the delay due to message length, $\frac{L}{W}$, is nearly equal to the delay due to the distance traveled, D. This minimum occurs at a surprisingly low dimension. For small networks, 1000 processors or less, the minimum latency is achieved with a two-dimensional network. Even for very large concurrent computers, networks with 4 or 6 dimensions are sufficient. In addition to providing low-latency, low-dimensional networks have several other advantages. They are easy to construct, since they fit into a plane with fewer *folds* than high-dimensional networks. Two dimensional networks are particularly easy to construct since they fit into the plane with no folds, and all channels are the same length. Low dimensional networks are also easier to interface to and control. Since they have fewer channels per node, they require less control logic to manage communications.

Virtual channels can be used to construct deadlock-free routing algorithms for all strongly connected interconnection networks including k-ary n-cubes. By making routing a function of the channel on which a message arrives at a node, and by multiplexing several virtual channels over a single physical channel, the cycles in a channel dependency graph can be broken into spirals to avoid deadlock. A virtual channel routing algorithm has the advantage that it can be used with *wormhole* [115] routing. In a wormhole network, flow control is performed on *flits* that cannot be interleaved. Conventional *structured buffer pool* deadlock avoidance algorithms are designed for *store-and-forward* networks, where flow control is performed at the level of packets that can be interleaved. These algorithms depend on the ability to interleave packets and thus cannot handle *wormhole* routing, since flits cannot be interleaved. The torus routing chip

(TRC), a self-timed VLSI chip, has been developed to demonstrate the feasibility of wormhole routing and virtual channels.

Low-latency processing elements are required to support fine-grain concurrent computation. A conventional processor executes about one hundred instructions to receive and interpret a single message. A message-driven processor directly interprets messages, eliminating this interpretation overhead. The instructions of a message-driven processor are messages. By performing automatic message reception and buffering, accelerating message lookup with an instruction translation lookaside buffer, and providing addressing modes for fast access to the context and receiver, the architecture of a message-driven processor is matched to the semantics of CST.

A machine with low-latency communications channels and processing elements is capable of supporting instruction-size granularity. In the past, concurrent computation has been performed with process-size granularity. With finer-grain concurrency, less memory is required at each processing node. Current machines require a large memory at each node to support a grain size large enough to keep their high latency from dominating computation time. Some argue that a large memory is required to store a copy of the operating system at each node; however, such a practice is wasteful. By properly layering the operating system, only a few bottom-level modules, e.g., method lookup and mail delivery, need to be replicated in every processing node. Higher level modules can be stored in a single processing node and cached in other nodes as required.

VLSI technology, being wire-limited, encourages specialization and locality. Storing data local to the logic that manipulates it results in shorter wires. A special-purpose VLSI chip has a fixed communication pattern and thus can make better use of the limited wires than a general-purpose chip that must support many different communication patterns. Unfortunately the high cost of designing a VLSI chip makes it impractical to build special-purpose VLSI chips for every application. However, specialization can be applied to many applications by building VLSI chips to accelerate operations on common classes of objects. These *object experts* can be shared among applications, offering performance comparable with special-purpose hardware while retaining much of the flexibility of a general-purpose machine.

Computer architecture encompasses the design of programming languages, data structures, and algorithms, as well as hardware. The approach taken here is to start with a programming paradigm, concurrent data structures, develop a notation, CST, write algorithms using this notation, and finally to organize hardware to support these algorithms. In contrast, many computer architects restrict themselves to the last step. They analyze existing algorithms and fine-tune architectures to execute these algorithms. The problem with this evolutionary approach is that it leads to inbreeding, amplifying both the good and bad features of existing computer architectures. The algorithms analyzed are optimized to run on the previous generation of machines, which were fine-tuned to execute the previous generation of algorithms, and so on. Each generation, algorithms are designed to make the best use of the good features of the machine and to avoid the bad. The next generation of machines, based on these algorithms, makes the good features better and ignores the bad since they were not frequently used by the algorithms. The worst effect of this approach to architecture is that it discourages new programming language features. Late-binding programming languages, for example, are often judged to be inefficient because they cannot be efficiently implemented on conventional machines. Late-binding languages are not inefficient; conventional architectures are just not well matched to these languages.

Powerful software features such as late-binding operators and automatic storage management that are often cited as inefficient need not be slow. These features can be made very efficient with a modest amount of hardware support. In fact, high-level features can lead to a more efficient computing system by replacing many ad hoc mechanisms with a single mechanism that can then be implemented in hardware. The key to a successful architecture is to identify a few simple mechanisms that can be accelerated by hardware.

A VLSI architecture must match the physical form of a machine to its logical function. Traditionally computer architects and designers have concentrated on the logical organization of machines, giving little consideration to their physical design. With VLSI technology this is no longer possible. VLSI technology is wire limited. To make best use of wiring resources, architects must carefully plan the physical design of their machines. For example, consider the interconnection networks analyzed in Chapter 5. Considering just the logical organization of the network, one quickly deduces (as Lang [81] did) that binary n-cubes offer superior performance because of their smaller *logical* diameter. When the physical implementation of the network is considered, however, one finds that in fact low-dimensional networks offer better performance because they make better use of their wires. The short logical diameter is no longer a great advantage since, after being embedded into a two- or three-dimensional implementation space, all network topologies have the same physical diameter.

Many experiments are required to refine the ideas presented in this thesis. A first step is to implement a compiler and run-time system to run CST on an existing concurrent machine such as the Caltech Cosmic Cube [114]. Because of the high latency of Cosmic Cube communication channels and the mismatch between CST semantics and the architecture of the Intel 8086-based processing nodes, such a system will be quite inefficient. Nevertheless this programming system will be used to gain practical experience in concurrent object-oriented programming and in building systems out of concurrent data structures.

The next step is to build hardware to improve the efficiency of the system. This is best done in stages.

1. Provide a low-latency communication facility by building a concurrent computer using the TRC for communications and a commercial microprocessor, such as the Motorola 68000 [94], for a processing element. Such a machine could be built in a relatively short time frame and would provide valuable experience in using a low-latency communication network.

2. Build a message-driven processor to complement the low latency of the TRC-based network. This machine will provide an efficient environment for fine-grain concurrent object-oriented programming and will provide further experience with this programming style.

3. Provide a powerful machine for demanding applications by constructing object experts for several commonly used classes such as floating point vectors and ordered sets.

The availability of a machine comparable to (3) above will stimulate much research on concurrent software. Concurrent operating systems will evolve to support fine-grain object-oriented programming. To run in a fine-grain machine with limited storage in each processing node, operating systems will be partitioned into layers with only the bottom layer duplicated in each node. Memory management functions will make the partitions between processing nodes invisible to user programs by maintaining a single name space across the machine. The system will relocate objects as required to make efficient use of memory and processing resources, dynamically balancing the load across the processing nodes. Systems will evolve to the point where a host is no longer required. Input/output devices will connect directly to processing nodes and will appear as objects to the system. Methods will be edited and compiled directly on the concurrent computer.

One fertile area for further research is the development of concurrent computer-aided-design (CAD) applications. The exponential growth in the complexity of VLSI systems that has made possible the construction of the machines described here has also exceeded the capacity of sequential CAD programs. For example, verification of a 10^5-transistor VLSI chip by logic simulation takes several weeks of CPU time. Since simulation time grows as the square of device complexity, one can project that a 10^6-transistor chip will require several years to verify. Concurrent CAD programs will give performance several orders of magnitude better than sequential applications, reducing verification time from years to days. More importantly, concurrent applications give performance that scales with the size of the problem. As VLSI chips become more complex, we will construct larger concurrent computers to design these chips. We will apply VLSI technology to solve the problem of VLSI complexity.

To exploit the low latency but high throughput of VLSI technology, we build concurrent computers consisting of many processing nodes connected by a network. Software is the real challenge in the development of these machines. It is difficult to focus the activity of large numbers of processing elements on the solution of a single problem. This thesis proposes a solution to the problem of programming concurrent computers: concurrent data structures. Most applications are built around data structures. The problem of coordinating the activity of many processing elements is solved once and encapsulated in a class definition for a concurrent data structure. This data structure is used to construct concurrent applications without further concern for the problems of communication and synchronization. The combination of VLSI and concurrency will make computers fast. The combination of object-oriented programming and concurrent data structures will make them easy to program.

Appendix A

Summary of Concurrent Smalltalk

Concurrent Smalltalk (CST) is an extension of the Smalltalk-80 programming language [53], [54], [76], [138] that incorporates *distributed objects*, concurrent message sending, and *locks*. The differences between CST and Smalltalk-80 are described in Chapter 2. This Appendix gives a brief summary of the entire programming language for those readers not familiar with Smalltalk-80. For a more complete description of the programming language, the interested reader should consult [53] or [138].

Classes

A CST program consists of a set of class declarations. Each class declaration describes the state and behavior of a class of objects and has the form shown in Figure A.1. The declaration contains the name of the class's superclass, specification of the class object, and specification of each instance of the class.

class: The class name identifies the class object, the object that contains the class variables and implements the class methods. The class object name is capitalized since the class is a global object[1] and by convention names of shared variables are capitalized.

---

[1]Smalltalk could be greatly improved by adding some type of scoping to class names so that a user could locally override a class in an application without changing the class used by the rest of the system.

class	<identifier>	*the class name*
superclass	<identifier>	*name of the superclass*
instance variables	[<identifier>]*	*state of each instance object*
class variables	[<identifier>]*	*state of the class object*
locks	[<identifier>]*	*locks controlling access to*
class methods		*each instance object*

    class methods ...

instance methods

    instance methods ...

Figure A.1: Class Declaration

---

superclass: The superclass name identifies the superclass from which the current class inherits variables and methods. The current class is declared as an extension of the superclass. All class and instance variables declared in the superclass are added to the lists specified in the class declaration. All class and instance methods that are not overridden in the class definition are also inherited from the superclass. The inheritance can extend through many levels of the superclass hierarchy, with the current class inheriting methods and variables from the superclass that were in turn inherited from the superclass' superclass, and so on.

instance variables: The private memory of each instance of the defined class. For example, if we define a class Point with instance variables x and y, then each instance of class Point is created with two local variables named x and y distinct from the variables in any other instance. The instance variables specified in this declaration are in addition to any instance variables specified by the superclass.

class variables: Variables shared by the class object and all instances of the class. There is only one instance of each class variable. This single copy of a class variables can be accessed by any instance of the class. Class variables are capitalized since they are shared variables.

locks: Locks are special instance variables that control concurrent access to objects.

class methods: Methods that define the behavior of the class object. Each method specifies a number of expressions to be performed in response to a message. Typically class methods handle tasks such as object creation.

instance methods: Methods that define the behavior of each instance of the class.

Messages

Everything in CST is done by passing messages. Sending a message to an object causes the object to execute one of its methods. A message has three parts:

receiver: The object to which the message is being sent.

selector: The type of message. The selector specifies the method the receiver is to execute.

arguments: Additional data required for the receiver to execute the method specified by the selector.

Here are some examples of message expressions.

```
theta sin
```

This message expression sends the message with selector sin and no arguments to theta, the receiver. A message like sin that has no arguments is called a *unary message*. In a unary message, the selector follows the receiver.

```
a + b
```

The receiver, a, is sent the message containing the selector, +, with argument, b. A message like + b, where there is a single argument and the selector consists of one or two special characters, is called a *binary message*.

```
foo at: 10 put: 'hello'
```

This keyword message sends the message with selector at:put: to object foo with arguments 10 and 'hello'. In keyword messages the selector consists of a keyword before each argument. Each keyword is terminated by a colon, ':'.

When an object receives a message, it looks up and executes the method that matches the message selector. The method lookup begins with the receiver first checking its own instance methods. If the method is not found in the receiver's class, the instance methods defined in the superclass are checked, and so on.

Literals

The receiver and arguments in a message expression may be variables, pseudo-variables, or literals. CST supports the following types of literals:

numbers: Numbers consist of an optional sign, an optional radix, an integer part, an optional fraction part, and an optional exponent. Here are some examples of numbers.

17	*an integer, radix 10 is default*
16rFF	*a radix 16 (hexadecimal) integer*
3.14159265358979	*pi*
-10.1e-2	*−0.101*
2r101e2	*2r10100 or 20*

characters: Character literals consist of a dollar sign, '$', followed by any character, e.g., $A.

strings: String literals consist of a sequence of characters delimited by single quotes, e.g., 'Hello World'. To insert a single quote into a string it is duplicated, e.g., 'don''t'.

symbols: Symbols or atoms consist of a hash mark followed by the name of the symbol, e.g., #slave.

arrays: A sequence of literals is denoted by the sequence with hash marks removed enclosed in parentheses, '()', and preceded by a hash mark, e.g., #(1 2 slave $A 'Element' 2r10001 (1 2 3)).

Assignment

To simplify assignment of values to variables, CST permits the result returned by a method to be assigned to a variable by using the backarrow, '←', character. For example, the message

   a ←3 + 2.

assigns to variable a the result of sending the message + 2 to the object 3. Assignment can be thought of as sending an at: variable put: expression message to the current environment.

Messages that do not include an assignment do not generate a reply. To wait for a message that returns no value, the message is preceded by a backarrow, '←', with no variable. For example, the message

   ←aRectangle display.

sends a display message to aRectangle and expects a reply from this message. The message

   aRectangle display.,

on the other hand, does not expect a reply.

Methods

An object's protocol[2] is defined by the instance methods in the class declaration. Two example method descriptions are shown in Figure A.2. The first method calculates the product of a sequence of integers beginning with the receiver and ending with the argument upperBound. This definition of the message rangeProduct: follows that of Theriault [128]. The second method is the contains method for class Interval described in Chapter 2 (Figure 2.3 on page 20). This method tests if a number is contained in a closed interval of numbers.

Each method description consists of the following parts:

---

[2]An object's protocol consists of the messages understood by an object.

---

instance methods for Integer
   rangeProduct: upperBound
     locks would go here
     | midPoint upperProd lowerProd |
     self = upperBound ifTrue: [
        ↑self]
     ifFalse: [
        midPoint ←self + upperbound // 2.
        lowerProd ←self rangeProduct: midPoint,
        upperProd ←midPoint rangeProduct: upperBound.
        ↑lowerProd * upperProd.]
instance methods for Interval
   contains: aNum   *tests for number in interval*
     require rwLock.
     | lin uin |
     lin ←l $\leq$ aNum,
     uin ←u $\geq$ aNum.
     ↑(lin and: uin)

Figure A.2: Methods

---

header: The method header consists of the selector that activates the method with pseudo-variables in place of arguments. When a message is received by an object, the object's method with the corresponding header is activated. Message arguments are bound to the pseudo-variables in the method header. Pseudo-variables are like instance variables except that they cannot be assigned to. For example, the header **rangeProduct: upper-Bound** specifies that the following method will be executed in response to a message with selector rangeProduct: and the pseudo-variable upperBound will be bound to the argument of the message.

concurrency control: An optional concurrency control line specifies a required set of locks, an excluded set of locks, and an optional escape expression. The method is allowed to execute only when no currently pending method requires an excluded lock or excludes a required lock. If the method is locked out, the escape expression is executed or, if no escape expression is present, the method is suspended. Because the contains: method requires rwLock and specifies no escape, it will be suspended if some previous method excluded rwLock and will be restarted only when all such methods have completed.

local variables: Local variables are declared between two vertical bars, '|'. For example, the rangeProduct: method declares three local variables, midPoint, upperProd, and lowerProd.

message expressions: The remainder of the method consists of message expressions. Messages are separated by commas, ',', or periods, '.'. A comma between two messages means that the second message can be sent before receiving a reply from the previous message. When a period follows a message, replies must be received from all previous messages whose results are assigned with a backarrow before the next message can be sent. For example, the rangeProduct: method sends messages to self and midPoint concurrently and then waits for replies from both messages before multiplying the two results.

Messages may be nested within other messages. The reply of one message, A, may specify the receiver or argument of another message, B. For example, the message

$$\text{self} = \text{upperBound ifTrue: } [\cdots] \text{ ifFalse: } [\cdots].$$

in method rangeProduct: first sends the message, = upperBound, to self and then sends the ifTrue:ifFalse: message to the reply of this first message. Three rules govern the parsing of these compound messages:

1. Any messages enclosed in parentheses, '()', are evaluated before the messages outside the parentheses.

2. Unary messages take precedence over binary messages, and binary messages take precedence over keyword messages.

3. For messages of equal precedence, evaluation proceeds from left to right.

Two special identifiers allow a method to refer to the receiver.

self is an expression that specifies the receiver.

super also specifies the receiver, but messages sent to super are interpreted by looking up the method beginning in the receiver's superclass. Messages to super are often used to inherit a method from the superclass while making additions in the subclass.

Messages return a value by preceding a message expression with an uparrow, '↑'. The value returned by the following message is in turn returned by the method. Preceding a variable by an uparrow returns the value of the variable. A downarrow, '↓', causes a method to terminate without returning a value.

Blocks

Like Smalltalk-80, CST has no built-in control structures. Instead, control structures are built by sending messages to blocks. Blocks are deferred sequences of message expressions that are executed when they are sent a value message. Blocks are like methods in that they have arguments, locks, and local variables; unlike methods, however, blocks may have free variables that are lexically scoped. That is, a block may refer to the local variables of the method in which it is defined. Here is an example block:

```
[:edge | require rwLock :var1 |
var1 ←edge flow.
var1 > 0 ifTrue: [↓]].
```

Blocks are enclosed in square brackets, '[]', and consist of the following parts.

argument list: The optional argument list specifies the names of pseudo-variables that are bound to arguments passed into the block with a value:

arg message. Each identifier in the list is preceded by a colon, ': '. For example, in the block above the pseudo-variable edge is an argument. If this block is sent the message value: anEdge, the block will be executed with pseudo-variable edge bound to object anEdge.

concurrency control: Like methods, blocks may optionally specify two sets of locks to control concurrent access to the block.

local variables: The optional variable list consists of a list of identifiers preceded by colons. Local variables exist only for one activation of the block. Each time a block receives a value message, it creates a new context with a new set of local variables, all initialized to nil.

message expressions: The remainder of the block consists of a sequence of message expressions. The sequence is interpreted as in a method except that uparrow, '↑', returns out of the method calling the block and downarrow, '↓', breaks out of the block or method calling the block. The last message expression in the block is the value of the block expression.

A block is activated by sending it a value message. When a block receives a value message, the arguments of the message are bound to the arguments of the block and the message expressions in the block are executed.

Distributed Objects

A distributed object is a collection of *constituent objects* (COs) that receive messages sent to the distributed object. Because a distributed object contains many independent constituents, it can process many messages simultaneously.

Distributed objects are declared as subclasses of class DistributedObject. A new distributed object is created by sending the newOn: message to the appropriate class object. For example, a new instance of a TallyCollection (described in Figure 2.1 on page 16) is created with the message

    aTallyCollection ← TallyCollection newOn: someNodes.

The argument of the newOn: message, someNodes, is a collection of processing nodes. The newOn: message creates a CO on each member of someNodes.

When a message is sent to a distributed object, it may be delivered to any constituent of that object[3]. It is possible to send a message to a specific constituent of a distributed object by indexing the object with the selector co:. For example, the message

> aTallyCollection tally: 'hello'.

is sent to any constituent of aTallyCollection. The message

> aTallyCollection co: 3 tally: 'hello'.

is sent to the third constituent of aTallyCollection. Constituents are indexed sequentially beginning with one. The pseudo-variables maxId, the total number of constituents, and myId, the index of self, are available to constituent objects for use in computing indices.

Common Messages

To describe all of the classes and messages in a Smalltalk system is beyond the scope of this appendix. I include the following list of common messages to assist the reader in understanding the CST code in this thesis. This list is by no means comprehensive.

Block

value This unary message causes a block with no arguments to be executed.

value: anObject $\cdots$ A block with i arguments is sent a message with i value: keywords, one for each argument. This message passes the arguments to the block and causes the block to execute.

whileTrue: aBlock A value message is repeatedly sent to the receiver. As long as the receiver replies with true, a value message is sent to aBlock, and the sequence is repeated. If the receiver replies with false, the method terminates.

whileFalse: aBlock This message is similar to whileTrue but with the receiver negated. As long as the receiver block evaluates to false, the argument block is iterated.

---

[3]One hopes that the mail system will be efficient and deliver the message to the nearest CO or perhaps the CO with the shortest message queue.

Boolean

ifTrue: aBlock Sends a value message to aBlock if the receiver is true.

ifFalse: aBlock Sends a value message to aBlock if the receiver is false.

ifTrue: trueBlock ifFalse: falseBlock Sends a value message to trueBlock if the receiver is true. Otherwise, if the receiver is false, a value message is sent to falseBlock.

ifFalse: falseBlock ifTrue: trueBlock Sends a value message to trueBlock if the receiver is true. Otherwise, if the receiver is false, a value message is sent to falseBlock.

Number

+ Addition.

- Subtraction.

* Multiplication.

/ Division.

// Integer division rounding to $-\infty$.

\\ Modulo (remainder of division after rounding to $-\infty$).

quo: Integer division rounding to 0.

rem: Modulo (remainder of division after rounding to 0).

abs Absolute value.

negated Additive inverse.

reciprocal Multiplicative inverse.

> The selectors abs, negated, and reciprocal are not terminated with a colon because they are unary messages. The selectors quo: and rem: are terminated by colons because they are keyword messages.

Appendix B

Unordered Sets

Many applications use unordered data structures and do not require the overhead necessary to support an ordered set concurrent data structure like the balanced cube of Chapter 3. In this appendix I present two unordered concurrent data structures. A *dictionary* can be used in applications that require a data structure to hold associations between objects but do not need to maintain an order relationship on the objects. A *union-find set* can be used in applications where sets of data are combined.

B.1 Dictionaries

A *dictionary* is a set of associations between pairs of objects. Each element of a dictionary is an ordered pair (aKey,anObject) that associates a key aKey with object anObject. A dictionary supports the following operations [1].

at: aKey   return the object associated with key aKey.

at: aKey put: anObject   add an object to the set.

delete: aKey   remove the object associated with key aKey from the set.

do: aBlock : send a value: anObject message to aBlock for each object in the set.

---

[1] The complete protocol of class Dictionary is given in Chapters 9 and 10 of [53]. Most of the protocol is omitted here for the sake of brevity.

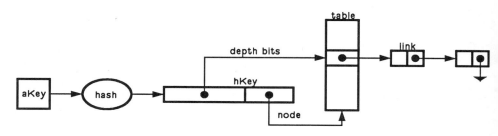

Figure B.1: A Concurrent Hash Table

Dictionaries represent binary relations. A common use of a dictionary is to represent the *name-of* relation by binding symbols to names. For example, the symbol table in a compiler is a dictionary.

Dictionaries can be implemented using a variety of data structures including radix search tries [109], binary search trees [2], and hash tables [109]. Hash tables are usually the structure of choice for sequential machines. The expected case access time for a hash table is $O(1)$ compared to $O(\log N)$ for the binary search tree, and hash tables make more efficient use of memory than radix search tries. In the past, one objection to hash tables was their fixed size; however, the recent development of extendible hashing [39], [85] makes hash tables efficient even for sets that change size dynamically.

Ellis has developed concurrent algorithms for extendible hashing [36] and linear hashing [35]. These algorithms involve locking schemes and protocols to support concurrent access to a shared hash table. Like most work on databases, this work assumes a disk-based system where multiple processes may compete for access to shared disk pages and is not directly applicable to concurrent computers.

Unlike most sequential data structures, the hash table is ideally suited for a concurrent implementation. The table is homogeneous and can be distributed uniformly over the nodes of a concurrent computer. Hash tables, unlike tree structures, have no root bottleneck.

A concurrent implementation of a hash table using a variant of bounded index hashing [85] is shown in Figure B.1. To see how this structure is used, consider the at: method for distributed object Hash Table shown in Figure B.2. Search key aKey is converted to a hashed key hKey by sending it the message hash.

The low-order bits of hKey are used to find the node that contains the data, while the next depth bits of hKey find the head of a linked list within the node. A linear search of this list is performed to return the object associated with aKey, or nil if this object is not found. The at:put: and delete: methods are obvious extensions of the at: method.

An extendible hash table [39] is implemented in each node. Each node's table is initialized to size 2^{depth}. When the number of entries increases beyond $\alpha 2^{depth}$ for some constant α, depth is incremented and the size of the table is doubled. The objects in the table need not be rehashed as the table grows. Doubling the size of the table simply increases by one the number of significant bits of hKey. The new entries in the table initially duplicate the old entries. As accesses are made to the table, the linked lists are split to shorten the access paths.

The do: aBlock method broadcasts aBlock to each node of the distributed hash table object. Each node enumerates the objects in its local table, sending each of them to aBlock. If aBlock updates no instance or method variables, it can be replicated, and the value methods can be processed in parallel. If aBlock updates instance or method variables, then the executions must be synchronized.

An operation on the table requires only two messages, a find:at: message to the node containing the key and the reply: message back. Thus, hashing is $O(1)$ in the number of elements in the set. However, since the destination of these messages is random, each message travels an average distance of $\frac{\log N}{2}$, where N is the number of nodes in the machine. This makes hash table access time grow $O(\log N)$ with the number of nodes in the machine.

Since the hash function randomizes access to a hash table, there is very little interaction between concurrent hash operations. Thus, the concurrency of hashing is $O(N)$.

B.2 Union-Find Sets

A *union-find* set, as the name implies, supports the operations of forming the union of two sets and finding the set to which an element belongs [2].

union: aSet   returns the union of the receiver with aSet. Both the receiver and
    aSet are modified to form the new set.

add: anElement   adds anElement to the receiver

---

[2]As with dictionary, this class supports a more complete protocol.

class	Hash Table	
superclass	Dictionary	*a distributed object*
instance variables	table	*table of links (key,data,next)*
	depth	*log of table size*
class variables		*none*
locks	rwLock	*implements readers and writers*

instance methods

   at: aKey   *find anObj in hash table*
      | hkey |
      hKey ←aKey hash.   *compute hashed key of object*
      (self at: (hkey mod maxId)) find: aKey at: (hKey/maxId).

private instance methods

   find: aKey at: hKey   *in proper node, find object*
      require rwLock
      | link |
      link ←table at: (hKey $\backslash\backslash 2^{depth}$).
      [link isNil] whileFalse: [
         (link key = aKey) ifTrue[requester reply: link data].
         link ←link next.]
      requester reply: nil.

Figure B.2: Concurrent Hashing

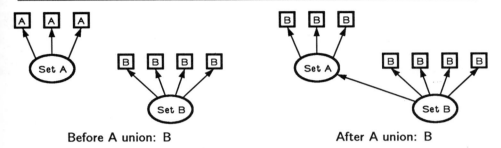

Before A union: B   After A union: B

Figure B.3: A Concurrent Union-Find Structure

set returns the set to which the receiver belongs. Elements of the set must support this message as well.

Algorithm 4.3 in [2] performs a sequence of union and find operations in time that is almost linear[3] in the number of operations performed, approximately constant time per operation. Unfortunately this algorithm has very poor concurrency. Every find requires traversing a tree from the leaves to the root. The root serializes finds since it can only process one message at a time.

To eliminate this root bottleneck, we store with each element the identity of the set to which the element belongs. As shown in Figure B.3, during a union operation the smaller set becomes a subset of the larger set. Each element of the smaller set must also be informed that it is now a member of the larger set. The code for these operations is shown in Figure B.4.

Only the elements of the smaller set are updated during a union operation. Since each time an element is updated the size of the set it belongs to has at least doubled, an element is updated at most $O(\log N)$ times, where N is the number of elements [4]. Thus, if we implement each of the sets with a dictionary or other constant access time structure, the average time per union operation will be $O(\log N)$. Find operations require $O(1)$ time. The concurrency of union-find operations depends on the balance of the resulting tree structure of sets.

---

[3]The time grows as $N\alpha(N)$ where α is the inverse of Ackerman's function and N is the number of operations.

[4]A similar approach is used in Section 4.6 of [2].

class	Union Find Set	
superclass	Set	*a distributed object*
instance variables	parent	*parent set if not self*
class variables		*none*
locks	meLock	
instance methods		

add: anObj   *add an element or subset to the set*
    require meLock
    ||
    anObj parent: self.
    ↑super add: anObj

union: aSet   *make smaller set a subset of larger*
    require meLock
    ||
    ((aSet size) > (self size)) ifTrue: [↑aSet union: self].
    ↑self add: aSet

private instance methods

parent: aSet   *inform all elements of new parent*
    require meLock exclude meLock
    ||
    parent ←aSet,
    self do: [:each | each parent: aSet].

Figure B.4: Concurrent Union-Find

Appendix C

On-Chip Wire Delay

Signal velocities on an integrated circuit are limited by the resistance and capacitance of the wire to be far less than the speed of light. Because the resistivity of integrated circuit wires is high, it is not possible to build good transmission lines on a chip. Instead, on-chip signal wires are lossy transmission lines with a delay proportional to the square of their length.

We can propagate a signal with linear delay by placing repeaters along a transmission line. Each repeater is an inverter of size S. The repeaters are spaced distance L apart. Let us make the following assumptions:

- The ratio of inverter input capacitance to transistor gate capacitance is X. For a CMOS inverter with the p-channel transistor twice the size of the n-channel transistor, $X = 3$.

- Transistors will be modeled by a linear resistance. The resistance of a minimum width transistor is R_t. The output resistance of each inverter is $R_{\text{inv}} = \frac{R_t}{S}$.

- The gate capacitance of a minimum width transistor is a constant, C_g, and scales linearly with device size. The input capacitance of each inverter is $C_{\text{inv}} = XSC_g$.

- The resistance of a unit length wire is $K_r R_t$. The resistance of the wire between two inverters is $R_w = LK_r R_t$.

- The capacitance of a unit length wire is $K_c C_g$. The capacitance of the wire between two inverters is $C_w = LK_c C_g$.

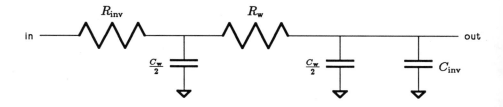

Figure C.1: Model of Inverter Driving Wire

We model one stage of the RC transmission line with repeaters with a Π network as shown in Figure C.1. Half of the distributed wire capacitance is lumped at each end of the wire. The output resistance of the driving repeater is added to the input end of the network, and the input capacitance of the receiving repeater is added to the output side of the network.

We approximate the delay between repeaters by multiplying each resistance by the capacitance it 'sees' and summing the products.

$$
\begin{aligned}
T \;&=\; R_{\mathrm{inv}}\left(C_w + C_{\mathrm{inv}}\right) + R_w\left(\frac{C_w}{2} + C_{\mathrm{inv}}\right), \\[2mm]
&=\; \frac{R_t}{S}\left(LK_cC_g + XSC_g\right) + LK_rR_t\left(\frac{LK_cC_g}{2} + XSC_g\right), \\[2mm]
&=\; R_tC_g\left(\frac{LK_c}{S} + X + \frac{L^2K_rK_c}{2} + LK_rXS\right).
\end{aligned}
\tag{C.1}
$$

To find the optimal repeater size, we take the partial derivative of T with respect to S

$$
\frac{\partial T}{\partial S} = R_tC_g\left(-\frac{LK_c}{S^2} + LK_rX\right).
\tag{C.2}
$$

Setting $\frac{\partial T}{\partial S}$ equal to zero and solving for S gives

$$S_{\text{opt}} = \sqrt{\frac{K_c}{XK_r}}.$$ (C.3)

To find the optimal repeater spacing, we take the partial derivative of $\frac{T}{L}$, the inverse of signal velocity, with respect to L

$$\frac{\partial \frac{T}{L}}{\partial L} = R_t C_g \left(-\frac{X}{L^2} + \frac{K_r K_c}{2} \right).$$ (C.4)

Setting this equal to zero and solving for L gives

$$L_{\text{opt}} = \sqrt{\frac{2X}{K_r K_c}}.$$ (C.5)

Substituting (C.3) and (C.5) back into (C.1) gives the delay of the optimal segment

$$T_{\text{opt}} = 2R_t C_g X \left(1 + \sqrt{2} \right).$$ (C.6)

Dividing (C.5) by (C.6) gives the maximum signal velocity in units of $\frac{\text{squares}}{\text{sec}}$

$$v_{\text{opt}} = \frac{\sqrt{2}}{2R_t C_g \sqrt{K_r K_c} X \left(1 + \sqrt{2} \right)}.$$ (C.7)

Let us put some real numbers into these equations. The following table gives approximate values for our four constants as a function of linear dimension, λ, in microns.

Parameter	Value	Units
R_t	10^4	Ω
C_g	4λ	fF
K_c	0.1	
K_r	$\frac{5 \times 10^{-6}}{\lambda}$	

For a 1μ technology $(\lambda = 0.5\mu)$, if we set $X = 3$, we can calculate:

- Optimal repeater size is $S \approx 60$.

- Optimal repeater spacing is $L \approx 2500$.

- Time between repeaters is $T \approx 300$ps.

- The maximum signal velocity, $v = \frac{L}{T} \approx 8 \times 10^6 \frac{m}{sec} \ll 3 \times 10^8 \frac{m}{sec}$.

This calculation has not been terribly accurate. Still, it is clear that signal velocities on integrated circuits are limited by resistance to be much less than the speed of light.

Glossary

acquaintance An object's, A's, acquaintances are those objects to which A can send messages. In most cases an object's acquaintances are its instance variables and class variables.

actor: A synonym for object.

algorithm: A finite set of instructions for solving a specific type of problem [73].

argument: An object passed as part of a message. Arguments are bound to pseudo-variables in the method executed in response to the method.

assignment: The process of binding an object to a variable. In CST assignment is indicated by a backarrow, '←'. For example, a ←b, assigns the value of b to variable a.

balanced cube: A concurrent ordered set data structure that maps the elements of an ordered set to the right subcubes of a binary n-cube.

balanced tree: An ordered set data structure based on a binary search tree whose height is kept within a constant factor of $\log_2 N$, where N is the number of data in the tree [74].

B-cube: A concurrent ordered set data structure where multiple data are stored in each node of a balanced cube.

B-tree: An ordered set data structure based on a tree with the following properties [74]:

1. Each internal node of a B-tree of order N has between N/2 and N children.

2. All leaves of a B-tree are at the same level and contain no data.

3. An internal node with k children contains k records.

4. The ith record of an internal node is greater than the $i - 1^{\text{st}}$ record.

5. All records stored in the ith child of a node, A, are greater than the $i - 1^{\text{st}}$ record stored in A and less than the ith record stored in A.

binary message: A message with a single argument and a selector composed of one or two special characters. For example, a + b and p $\leq$ q are binary messages.

binary n-cube: An interconnection topology with $N = 2^n$ nodes where each node has a binary address, a, and is connected to those nodes whose addresses differ from a in exactly one bit position, $a \oplus 2^i$, $0 \leq i < n$.

binding: The process of associating meaning with an object. For example, object-oriented programming languages bind meaning to message objects by associating a method with each message.

block: In Concurrent Smalltalk, a block is a sequence of deferred message expressions along with arguments, locks and local variables. A block is executed when it receives a value message.

[:each | require rwLock :var1 :var2 | message1. message2]

The block above, for example, has a single argument, each, requires a lock, rwLock, and has two variables, var1 and var2. When it receives a value: arg message, this block binds each to arg and executes the two messages.

cache: A small, fast memory used to hold frequently accessed data.

class: An object that describes the state and behavior of objects of a certain type.

class variable: A variable shared by objects of a certain class. It can be accessed by the class itself and by any instance of the class.

communication channel: The hardware used to transmit information between the nodes of a network. The channel includes the physical wires that carry the information, the buffers or queues that store information in transit, and the logic that controls information flow.

computer architecture: The process of organizing a computer system to apply available technology to the solution of a set of problems.

concurrent algorithm: An algorithm for a concurrent computer.

concurrent computer: A computer composed of many autonomous processing elements connected by a network. The term concurrent is used rather
than the term parallel to emphasize the autonomous nature of the processing elements [113].

concurrent data structure: A data structure that can perform many operations simultaneously.

constituent object (CO): An object that is part of a distributed object.
Constituent objects receive messages sent to the distributed object.

data abstraction: Data abstraction separates an object's protocol, the messages an object understands, from an object's implementation, how the
object responds to the messages in its protocol.

data structure: A collection of data on which some relations are defined.

deadlock: Deadlock occurs when no progress can be made because of a cyclic
conflict for resources. In an interconnection network deadlock occurs when
no message can advance toward its destination because the queues of the
message system are full.

degree: The degree of a vertex, v, is the number of edges incident on v.

diameter: The maximum over all pairs of vertices of the length of the shortest
path between two vertices in a graph.

direct network: An interconnection network in which the terminal nodes are
also the switching elements as opposed to an indirect network in which
the terminals and switching elements are distinct.

distributed object: An object consisting of a collection of constituent objects. A message sent to the distributed object may be received by any
constituent of the object.

edge: An ordered pair of vertices.

ensemble machine: A machine consisting of an ensemble of processing nodes
connected by a network [112]. The processing nodes of an ensemble machine may be autonomous as in a concurrent computer, or they may
operate in lockstep as in a SIMD [44] parallel computer.

flit: A FLow control digIT, the smallest unit of information that can be accepted or refused by a communication channel or queue. One or more flits
make up a packet. Individual flits do not contain sequencing or routing
information and thus flits in a packet cannot be interleaved with flits of
another packet.

heap: A data structure for implementing a priority queue. A heap is organized as a binary tree with one record stored in each node of the tree. The tree is ordered so that the record stored in each node is greater than the records stored in both of its children.

hypercube: A k-ary n cube with dimension, n, greater than three. Hypercube is often incorrectly used as a synonym for binary n-cube; however, the radix of a hypercube is not restricted to be two.

identifier: A name or symbol. In CST an identifier consists of a letter possibly followed by a sequence of letters and digits.

inheritance: In an object-oriented language, a subclass *inherits* behavior from its superclass.

instance: An instance of a class, A, is an object of class A.

instance variable: A variable local to a particular instance of an object. Instance variables make up an object's private memory.

interconnection network: A communication network used to connect the processing nodes of an ensemble machine.

indirect network: An interconnection network in which the terminal nodes are distinct from the switching elements as opposed to a direct network in which the terminals contain the switching elements.

k-ary n-cube: An interconnection topology with $N = k^n$ nodes. Each node in a k-ary n-cube has an n-digit radix k address, $a = a_{n-1}, \ldots, a_0$, and is adjacent to those nodes with addresses $b = b_{n-1}, \ldots, b_0$ that differ from a in only one digit, say the jth, and this digit differs only by one, $a_j = b_j \pm 1$. Binary n-cubes are a special case of k-ary n-cubes where $k = 2$.

keyword message: A message consisting of a selector and one or more arguments where the selector is a sequence of keywords terminated with colons, ':', one preceding each argument. For example, the message receiver at: 8 put: 'arg2' is a keyword message with selector at:put: and arguments 8 and 'arg2'.

late binding: Binding meaning to objects as late as possible, usually at runtime. In contrast, early binding usually takes place at compile time.

latency: The elapsed time required to perform an operation. The latency of a message transmission is the elapsed time from the time the first flit of the message leaves the source to the time the last flit of the message arrives at the destination.

lock: A programming construct used to restrict concurrent access to an object.

message: In an object-oriented programming language, a message is a request for an object to perform some action. Messages consist of three parts: a receiver that specifies the object which is to receive the message, a selector that specifies the type of action to be performed, and arguments that supply additional information required to perform the action. In an interconnection network, a message is a logical unit of communication. A message may be broken down into a number of packets, physical units of communication that contain routing and sequencing information. Packets in turn may be broken down into flits.

message-passing concurrent computer: A concurrent computer in which the processing nodes communicate by passing messages over communication channels.

method: A description of how an object is to respond to a message. Methods in object-oriented programming languages are similar to procedures and subroutines in conventional programming languages.

multiprogrammed system: A computer system that supports multiple processes on a single processor.

object: The primitive element of an object-oriented programming system. An object consists of a state and a behavior. The state of an object is made up of a number of variables or acquaintances. The behavior of an object is specified by a number of methods. The object executes these methods in response to particular messages.

object expert: A processing element specialized to operate on a restricted class of objects. An object expert contains both storage for instances of this class of objects and logic specialized to operate on these objects.

packet: In a communication network a packet is the smallest unit of information that contains routing information. Packets may be broken down into flits.

path: A sequence of connected edges in a graph.

protocol: The set of messages that an object understands.

receiver: The object to which a message is sent.

selector: A part of a message specifying the type of operation to be performed by the object receiving the message.

self-timed: A design discipline where the sequencing of events is controlled by the internal delays of elements rather than by an external clock.

sequential computer: A computer that executes instructions one at a time.

shared-memory concurrent computer: A concurrent computer in which the processing elements communicate by reading and writing shared storage locations.

store-and-forward routing: A routing strategy where an entire packet is stored in each node along a multi-hop path before transmission to the next node is initiated.

strongly connected: A graph is strongly connected if there exists a path from every vertex in the graph to every other vertex.

structured buffer pool: A technique used to prevent deadlock in an interconnection network by controlling the allocation of buffers to packets.

subclass: A class that inherits methods and variables from an existing class, its superclass.

superclass: The class from which methods and variables are inherited.

throughput: The total number of operations performed per unit time.

tori: Plural of torus.

torus: Topologically, a torus is a doughnut shaped surface. In terms of interconnection networks, torus is a synonym for k-ary n-cube.

tree: In Computer Science a tree refers to a hierarchical data structure organized as a connected acyclic directed graph where the in-degree of each vertex is less than or equal to one.

useful: In a flow graph, an edge, e, is useful from vertex u to vertex v, denoted useful(u,v) if $e = (u, v)$ and $f(e) < c(e)$, or $e = (v, u)$ and $f(e) > 0$.

vertex: A part of a graph.

virtual channels: A technique for preventing deadlock in an interconnection network by multiplexing several *virtual channels*, each with its own queue, over a single physical channel and restricting the routing on virtual channels so that there are no cyclic dependencies amongst channels.

very large scale integration (VLSI): A technology for fabricating integrated circuits containing over 10^4 devices.

wafer scale integration (WSI): A technology for fabricating integrated circuits the size of wafers (50-150mm on a side).

wormhole routing: A routing strategy where each flit of a packet is immediately forwarded to the next node along a multi-hop path without waiting for the rest of a packet to arrive.

Bibliography

[1] Agha, Gul A., *Actors: A Model of Concurrent Computation in Distributed Systems*, MIT Artificial Intelligence Laboratory, Technical Report 844, June 1985.

[2] Aho, Alfred V., Hopcroft, John E., and Ullman, Jeffrey D., *The Design and Analysis of Computer Algorithms*, Addison-Wesley, Reading, Mass., 1974.

[3] Athas, W.C., *XCPL, an Experimental Concurrent Language*, Dept. of Computer Science, California Institute of Technology, Technical Report 5196, 1985.

[4] Backus, John, "Can Programming Be Liberated from the von Neumann Style? A Functional Style and Its Algebra of Programs," *CACM*, Vol. 21, No. 8, August 1978, pp. 613-641.

[5] Baird, Henry S., "Fast Algorithms for LSI Artwork Analysis," *Proceedings, 14th ACM/IEEE Design Automation Conference,* 1977, pp. 303-311.

[6] Barnes, Earl R., "An Algorithm for Partitioning the Nodes of a Graph," *SIAM J. Alg. Disc. Meth.,* Vol. 3, No. 4, December 1982, pp. 541-550.

[7] Batcher, K.E., "Sorting Networks and Their Applications," *Proceedings AFIPS FJCC,* Vol. 32, 1968, pp. 307-314.

[8] Batcher, K.E., "The Flip Network in STARAN," *Proceedings, 1976 International Conference on Parallel Processing,* pp. 65-71.

[9] Baudet, Gerard M., *The Design and Analysis of Algorithms for Asynchronous Multiprocessors,* Ph.D. Thesis, Department of Computer Science Carnegie-Mellon University, Technical Report CMU-CS-78-116, 1978.

[10] Benes, V.E., *Mathematical Theory of Connecting Networks and Telephone Traffic,* Academic, New York, 1965.

[11] Birtwhistle, Graham M., Dahl, Ole-Johan, Myhrhaug, Bjorn, and Nygaard, Kristen, *Simula Begin,* Petrocelli, New York, 1973.

[12] Blodgett, A.J. and Barbour, D.R., "Thermal Conduction Module: A High Performance Multilayer Ceramic Package," *IBM J. of Research and Development,* Vol. 26, No. 1, January 1982, pp. 30-36.

[13] Browning, Sally, *The Tree Machine: A Highly Concurrent Computing Environment,* Dept. of Computer Science, California Institute of Technology, Technical Report 3760, 1985.

[14] Bryant, R., "A Switch-Level Model and Simulator for MOS Digital Systems," *IEEE Transactions on Computers,* Vol. C-33, No. 2, February 1984, pp. 160-177.

[15] Chandy, K.M. and Misra, J., "Distributed Computation on Graphs: Shortest Path Algorithms," *CACM,* Vol. 25, No. 11, November 1982, pp. 833-837.

[16] Chapman, P.T. and Clark K., Jr., "The Scan-Line Approach to Design Rules Checking," *Proceedings, 21ˢᵗ ACM/IEEE Design Automation Conference,* 1984, pp. 235-241.

[17] Clinger, W.D., *Foundations of Actor Semantics,* MIT Artificial Intelligence Laboratory, Technical Report 633, May 1981.

[18] Condon, Joseph H. and Thompson, Ken, "Belle Chess Hardware," *Advances in Computer Chess,* Vol. 3, Pergamon Press, Oxford, 1982, pp. 45-54.

[19] Dahl, O.J. and Nygaard, K., "SIMULA - An Algol-Based Simulation Language," *CACM,* Vol. 9, No. 9, September 1966, pp. 671-678.

[20] Dally, William J. and Seitz, Charles L., *The Balanced Cube: A Concurrent Data Structure,* Dept. of Computer Science, California Institute of Technology, Technical Report 5174:TR:85, February 1985, early release of [21].

[21] Dally, William J. and Seitz, Charles L., *The Balanced Cube: A Concurrent Data Structure,* Dept. of Computer Science, California Institute of Technology, Technical Report 5174:TR:85, May 1985.

[22] Dally, William J. and Kajiya, J., "An Object Oriented Architecture," *Proceedings, 12th International Symposium on Computer Architecture,* 1985, pp. 154-161.

[23] Dally, William J. and Bryant, Randal E., "A Hardware Architecture for Switch-Level Simulation" *IEEE Transactions on Computer-Aided Design,* Vol. CAD-4, No. 3, July 1985, pp. 239-250.

[24] Dally, William J. and Seitz, Charles L., *Deadlock-Free Message Routing in Multiprocessor Interconnection Networks,* Dept. of Computer Science, California Institute of Technology, Technical Report 5206:TR:86, 1986.

[25] Dally, William J. and Seitz, Charles L., "The Torus Routing Chip," *J. Distributed Systems,* Vol. 1, No. 3, 1986, pp. 187-196.

[26] Dally, William J., *CNTK: An Embedded Language for Circuit Description,* Dept. of Computer Science, California Institute of Technology, Display File, in preparation.

[27] Dally, William J., "Wire-Efficient VLSI Multiprocessor Communication Networks," *1987 Stanford Conference on Advanced Research in VLSI,* MIT Press, Cambridge, MA, 1987, pp. 391-415.

[28] Dally, William J., et.al., "Architecture of a Message-Driven Processor," to appear in *Proceedings, 14th International Symposium on Computer Architecture,* 1987.

[29] Dijkstra, E.W., "A note on two problems in connexion with graphs," *Numerische Mathematik,* Vol. 1, 1959, pp. 269-271.

[30] Dijkstra, E.W. and Scholten, C.S., "Termination Detection for Diffusing Computations," *Information Processing Letters,* Vol. 11, No. 1, August 1980, pp. 1-4.

[31] Donath, W.E. and Wong, C.K., "An Efficient Algorithm for Boolean Mask Operations," *Proceedings, 20th ACM/IEEE Design Automation Conference,* 1983, pp. 358-360.

[32] Edmonds, J. and Karp, R.M., "Theoretical Improvements in Algorithmic Efficiency for Network Flow Problems," *JACM,* Vol. 19, No. 2, April 1972, pp. 248-264.

[33] Ellis, C.S., "Concurrent Search and Insertion in AVL Trees," *IEEE Transactions on Computers,* Vol. C-29, No. 9, September 1980, pp. 811-817.

[34] Ellis, C.S., "Concurrent Search and Insertion in 2-3 Trees," *Acta Informatica,* Vol. 14, 1980, pp. 63-86.

[35] Ellis, C.S., *Concurrency and Linear Hashing,* Computer Science Department, University of Rochester, TR 151, March 1985.

[36] Ellis, C.S., *Distributed Data Structures, A Case Study,* Computer Science Department, University of Rochester, TR 150, August 1985.

[37] Even, S. and Tarjan, R.E., "Network Flow and Testing Graph Connectivity," *SIAM J. Computing,* Vol. 4, 1975, pp. 507-518.

[38] Even, Shimon, *Graph Algorithms,* Computer Science Press, Rockville, Md., 1979.

[39] Fagin, Ronald, Nievergelt, Jurg, Pippenger, Nicholas and Strong, H. Raymond, "Extendible Hashing- A Fast Access Method for Dynamic Files," *ACM Transactions on Database Systems,* Vol. 4, No. 3, September 1979, pp. 315-344.

[40] Fiduccia, C.M. and Mattheyses R.M., "A Linear-Time Heuristic for Improving Network Partitions," *Proceedings, 19th ACM/IEEE Design Automation Conference,* 1982, pp. 175-181.

[41] Filman, Robert E. and Friedman, Daniel P., *Coordinated Computing, Tools and Techniques for Distributed Software,* McGraw-Hill, New York, 1984, Ch. 17.

[42] Fisher, A.L. and Kung, H.T., "Synchronizing Large VLSI Processor Arrays," *IEEE Transactions on Computers,* Vol. C-34, No. 8, August 1985, pp. 734-740.

[43] Floyd, R.W., "Algorithm 97: Shortest Path," *CACM,* Vol. 5, No. 6, June 1962, p. 345.

[44] Flynn, Michael J., "Some Computer Organizations and Their Effectiveness," *IEEE Transactions on Computers,* Vol. C-21, No. 9, September 1972.

[45] Ford, L.R., Jr. and Fulkerson, D.R., *Flows in Networks,* Princeton University Press, Princeton, N.J., 1962.

[46] Galil, Z. and Naamad, A., "Network Flow and Generalized Path Compression," *Proceedings, 11th ACM Symposium on the Theory of Computing,* 1979, pp. 13-26.

[47] Galil, Z., "An $O(V^{\frac{5}{3}}E^{\frac{2}{3}})$ Algorithm for the Maximal Flow Problem," *Acta Informatica*, Vol. 14, 1980, pp. 221-242.

[48] Galil, Z. "On the Theoretical Efficiency of Various Network Flow Algorithms," *Theoretical Computer Science*, Vol. 14, 1981, pp. 103-111.

[49] Garey, M.R. and Johnson D.S., *Computers and Intractibility, A Guide to the Theory of NP-Completeness*, W. H. Freeman and Company, 1979, p. 209.

[50] Gelernter, David, "A DAG-Based Algorithm for Prevention of Store-and-Forward Deadlock in Packet Networks," *IEEE Transactions on Computers*, Vol. C-30, No. 10, October 1981, pp. 709-715.

[51] Gerla, Mario, and Kleinrock, Leonard, "Flow Control: A Comparative Survey," *IEEE Transactions on Communications*, Vol. COM-28, No. 4, April 1980, pp. 553-574.

[52] Glasser, Lance A. and Dobberpuhl, Daniel W., *The Design and Analysis of VLSI Circuits*, Addison-Wesley, Reading, Mass., 1985.

[53] Goldberg, Adele and Robson, David, *Smalltalk-80: The Language and its Implementation*, Addison-Wesley, Reading, Mass., 1983.

[54] Goldberg, Adele, *Smalltalk-80: The Interactive Programming Environment*, Addison-Wesley, Reading, Mass., 1984.

[55] Goodman, J., "Using Cache Memories to Reduce Processor-Memory Traffic," 10th *Annual Symposium on Computer Architecture*, June 1983.

[56] Gottlieb, Alan, et al., "The NYU Ultracomputer - Designing an MIMD Shared Memory Parallel Computer," *IEEE Transactions on Computers*, Vol. C-32, No. 2, February 1983, pp. 175-189.

[57] Gottlieb, Alan, et al., "Basic Techniques for the Efficient Coordination of Very Large Numbers of Cooperating Sequential Processors," *ACM TOPLAS*, Vol. 5, No. 2, April 1983, pp. 164-189.

[58] Gray, H.J. and Levonian P.V., "An Analog-to-Digital Converter for Serial Computing Machines," *Proceedings of the I.R.E.*, Vol. 41, No.10, October 1953, pp.1462-1465.

[59] Guibas, L.J., Kung, H.T., and Thompson, C.D., "Direct VLSI Implementation of Combinatorial Algorithms," *Proceedings, Caltech Conference on VLSI*, 1979, pp. 509-525.

[60] Gunther, Klaus D., "Prevention of Deadlocks in Packet-Switched Data Transport Systems," *IEEE Transactions on Communications,* Vol. COM-29, No. 4, April 1981, pp. 512-524.

[61] Hewitt, Carl, "The Apiary Network Architecture for Knowledgeable Systems," *Conference Record of the 1980 LISP Conference,* 1980, pp. 107-117.

[62] Hill, F.J. and Peterson, G.R., *Digital Systems: Hardware Organization and Design,* Wiley, New York, 1978.

[63] Hillis, W. Daniel., *The Connection Machine (Computer Architecture for the New Wave),* MIT Artificial Intelligence Laboratory, AI Memo No. 646, September 1981.

[64] Hoare, C.A.R., "Communicating Sequential Processes," *CACM,* Vol. 21, No. 8, August 1978, pp. 666-677.

[65] Hu, T.C., *Combinatorial Algorithms,* Addison-Wesley, 1982.

[66] Inmos Limited, *IMS T424 Reference Manual,* Order No. 72 TRN 006 00, Bristol, United Kingdom, November 1984.

[67] Intel Scientific Computers, *iPSC User's Guide,* Order No. 175455-001, Santa Clara, Calif., Aug. 1985.

[68] Kermani, Parviz and Kleinrock, Leonard, "Virtual Cut-Through: A New Computer Communication Switching Technique," *Computer Networks,* Vol 3., 1979, pp. 267-286.

[69] Kernighan, B.W. and Lin, S., "An Efficient Heuristic Procedure for Partitioning Graphs," *Bell System Technical Journal,* February 1970, pp. 291-307.

[70] Kernighan, B.W. and Ritchie, D., *The C Programming Language,* Prentice-Hall, Englewood Cliffs, N.J., 1978.

[71] Kirkpatrick, S., Gelatt, C.D. Jr., Vecchi, M.P., "Optimization by Simulated Annealing," *Science,* Vol. 220, No. 4598, 13 May 1983, pp. 671-680.

[72] Kleinrock, Leonard, *Queueing Systems, Volume 2: Computer Applications,* Wiley, New York, 1976, pp. 438-440.

[73] Knuth, Donald E., *The Art of Computer Programming, Volume 1/ Fundamental Algorithms,* Addison-Wesley, Reading, Mass., 1973.

[74] Knuth, Donald E., *The Art of Computer Programming, Volume 3/ Sorting and Searching*, Addison-Wesley, Reading, Mass., 1973.

[75] Knuth, Donald E. *The TEXbook*, Addison-Wesley, Reading, Mass., 1984.

[76] Krasner, Glenn, *Smalltalk-80: Bits of History, Words of Advice*, Addison-Wesley, Reading, Mass., 1983.

[77] Kung, H.T., "The Structure of Parallel Algorithms," *Advances in Computers*, Vol. 19, 1980, pp. 65-112.

[78] Kung, H.T. and Lehman, P.L., "Concurrent Manipulation of Binary Search Trees," *ACM Transactions on Database Systems*, Vol. 5, No. 3, September 1980, pp. 354-382.

[79] Kyocera, Inc., *Design Guidelines, Multilayer Ceramics*, CAT/2T8403TM.

[80] Lamport, Leslie, *The LaTEX Document Preparation System*, Second Preliminary Edition, 1983.

[81] Lang, C.R. Jr., *The Extension of Object-Oriented Languages to a Homogeneous, Concurrent Architecture*, Dept. of Computer Science, California Institute of Technology, Technical Report 5014, May 1982.

[82] Lawrie, Duncan H., "Alignment and Access of Data in an Array Processor," *IEEE Transactions on Computers*, Vol. C-24, No. 12, December 1975, pp. 1145-1155.

[83] Lehman, P.L. and Yao, S.B., "Efficient Locking for Concurrent Operations on B-Trees," *ACM Transactions on Database Systems*, Vol. 6, No. 4, December 1981, pp. 650-670.

[84] Levitt, K.N. and Kautz, W.H., "Cellular Arrays for the Solution of Graph Problems," *CACM*, Vol. 15, No. 9, September 1972, pp. 789-801.

[85] Lomet, David B., "Bounded Index Exponential Hashing," *ACM Transactions on Database Systems*, Vol. 8, No. 1, March 1983, pp. 136-165.

[86] Malhotra, V.M., Kumar, M.P., and Maheshwari, S.N., "An $O(|V|^3)$ Algorithm for Finding Maximum Flows in Networks," *Information Processing Letters*, Vol. 7, No. 6, October 1978, pp. 277-278.

[87] Marberg, J.M. and Gafni, E., "An $O(N^3)$ Distributed Max-Flow Algorithm," *Proceedings, 18th Princeton Conference on Information Sciences and Systems,* 1984, pp. 478-482.

[88] Mead, Carver A. and Conway, Lynn A., *Introduction to VLSI Systems,* Addison-Wesley, Reading, Mass., 1980.

[89] Mead, Carver A. and Rem, Martin, "Cost and Performance of VLSI Computing Structures," *IEEE J. Solid-State Circuits,* Vol. SC-14, No. 2, April 1979, pp. 455-462.

[90] Mead, Carver A. and Rem, Martin, "Minimum Propagation Delays in VLSI," *IEEE J. Solid-State Circuits,* Vol. SC-17, No. 4, August 1982, pp. 773-775.

[91] Merlin, Philip M. and Schweitzer, Paul J., "Deadlock Avoidance in Store-and-Forward Networks-I: Store-and-Forward Deadlock," *IEEE Transactions on Communications,* Vol. COM-28, No. 3, March 1980, pp. 345-354.

[92] Miklosko, J. and Kotov, V.E., *Algorithms, Software and Hardware of Parallel Computers,* VEDA, Publishing House of the Slovak Academy of Sciences, Bratislava, 1984.

[93] Moore, Gordon, "VLSI: Some Fundamental Challenges," *IEEE Spectrum,* April 1979, pp. 30-37.

[94] Motorola Inc., *MC68000 16-bit Microprocessor User's Manual,* Third Edition, Prentice Hall, Englewood Cliffs, N.J., 1982.

[95] Ousterhout, John K., "Corner Stitching: A Data-Structuring Technique for VLSI Layout Tools," *IEEE Transactions on Computer Aided Design,* Vol. CAD-3, No. 1, January 1984, pp. 87-100.

[96] Ousterhout, John K., et al., "The Magic VLSI Layout System," *IEEE Design and Test of Computers,* Vol. 2, No. 1, February 1985, pp. 19-30.

[97] Papadimitriou, C.H. and Steiglitz, K., *Combinatorial Optimization: Algorithms and Complexity,* Prentice Hall, 1982.

[98] Pease, M.C., III, "The Indirect Binary n-Cube Microprocessor Array," *IEEE Transactions on Computers,* Vol. C-26, No. 5, May 1977, pp. 458-473.

[99] Peltzer, Douglas L., "Wafer-Scale Integration: The Limits of VLSI?" *VLSI Design,* September 1983, pp. 43-47.

[100] Peterson, James L., "Petri Nets," *Computing Surveys,* Vol. 9, No. 3, September 1977, pp. 223-252.

[101] Pfister, G.F., "The Yorktown Simulation Engine: Introduction," *Proceedings, 19th ACM/IEEE Design Automation Conference,* 1982, pp. 51-54.

[102] Pfister, G.F., et al., "The IBM Research Parallel Processor Prototype (RP3): Introduction and Architecture," *IEEE 1985 Conf. on Parallel Processing,* August, 1985, pp. 764-771.

[103] Pfister, G.F. and Norton, V.A., "Hot Spot Contention and Combining in Multistage Interconnection Networks," *IEEE Transactions on Computers,* Vol. C-34, No. 10, October 1985, pp. 943-948.

[104] Quinn, Michael J. and Narsingh, Deo, "Parallel Graph Algorithms," *Computing Surveys,* Vol. 16, No. 3, September 1984, pp. 319-348.

[105] Quinn, Michael J. and Yoo, Year Back, "Data Structures for the Efficient Solution of Graph Theoretic Problems on Tightly-Coupled MIMD Computers," *Proceedings, 1984 International Conference on Parallel Processing,* 1984, pp. 431-438.

[106] Ramamoorthy, C.V. and Li, H.F., "Pipeline Architecture," *ACM Computing Surveys,* Vol. 9, No. 1, March 1977, pp. 61-102.

[107] Russo, R.L., Oden, P.H., and Wolff, P.K., "A Heuristic Procedure for the Partitioning and Mapping of Computer Logic Blocks to Modules," *IEEE Transactions on Computers,* Vol. C-20, 1971, pp. 1455-1462.

[108] Schwartz, J.T., "Ultracomputers," *ACM TOPLAS,* Vol. 2, No. 4, October 1980, pp. 484-521.

[109] Sedgewick, Robert, *Algorithms,* Addison-Wesley, Reading, Mass., 1983.

[110] Seigel, Howard Jay, "Interconnection Networks for SIMD Machines," *IEEE Computer,* Vol. 12, No. 6, June 1979, pp. 57-65.

[111] Seitz, Charles L., "System Timing" in *Introduction to VLSI Systems,* C. A. Mead and L. A. Conway, Addison-Wesley, 1980, Ch. 7.

[112] Seitz, Charles L., *Experiments with VLSI Ensemble Machines,* Dept. of Computer Science, California Institute of Technology, Technical Report 5102, October 1983.

[113] Seitz, Charles L., "Concurrent VLSI Architectures," *IEEE Transactions on Computers,* Vol. C-33, No. 12, December 1984, pp. 1247-1265.

[114] Seitz, Charles L., "The Cosmic Cube," *CACM,* Vol. 28, No. 1, Jan. 1985, pp. 22-33.

[115] Seitz, Charles L., et al., *The Hypercube Communications Chip,* Dept. of Computer Science, California Institute of Technology, Display File 5182:DF:85, March 1985.

[116] Seitz, Charles L., et al., "Hot-Clock nMOS," *1985 Chapel Hill Conference on Very Large Scale Integration,* Henry Fuchs, ed., Computer Science Press, Rockville, Md., 1985.

[117] Seraphim, D.P. and Feinberg, I., "Electronic Packaging Evolution in IBM," *IBM J. of Research and Development,* Vol. 25, No. 5, September 1981, pp. 617-629.

[118] Shiloach, Y. and Vishkin, U., "An $O(n^2 \log n)$ Parallel MAX-FLOW Algorithm," *J. Algorithms,* Vol. 3, No. 2, June 1982, pp. 128-146.

[119] Siewiorek, D.P., Bell, C.G., and Newell, A., *Computer Structures: Principles and Examples,* McGraw-Hill, New York, 1982.

[120] Sleator, Daniel D.K., *An $O(nm \log n)$ Algorithm for Maximum Network Flow,* Ph.D. Thesis, Department of Computer Science, Stanford University, Report No. STAN-CS-80-831, December 1980.

[121] Spira, P.M., "A New Algorithm for Finding All Shortest Paths in a Graph of Positive Arcs in Average Time $O(n^2 \log^2 n)$," *SIAM J. Computing,* Vol. 2, No. 1, pp. 28-32.

[122] Steele, Craig S., *Placement of Communicating Processes on Multiprocessor Networks,* Dept. of Computer Science, California Institute of Technology, Technical Report 5184, 1985.

[123] Stefik, Mark and Bobrow, Daniel G., "Object-Oriented Programming: Themes and Variations," *AI Magazine,* Vol. 6, No. 4, Winter 1986, pp. 40-62.

[124] Stone, H.S., "Parallel Processing with the Perfect Shuffle," *IEEE Transactions on Computers,* Vol. C-20, No. 2, February 1971, pp. 153-161.

[125] Su, Wen-King, Faucette, Reese, and Seitz, Charles L., *C Programmer's Guide to the Cosmic Cube,* Dept. of Computer Science, California Institute of Technology, Technical Report 5203, September 1985.

[126] Sullivan, H. and Bashkow, T.R., "A Large Scale Homogeneous Machine," *Proc. 4th Annual Symposium on Computer Architecture,* 1977, pp. 105-124.

[127] Tanenbaum, A. S., *Computer Networks,* Prentice Hall, Englewood Cliffs, N.J., 1981.

[128] Theriault D.G., *Issues in the Design and Implementation of Act2,* MIT Artificial Intelligence Laboratory, Technical Report 728, June 1983.

[129] Thompson, C.D., *A Complexity Theory of VLSI,* Department of Computer Science, Carnegie-Mellon University, Technical Report CMU-CS-80-140, August 1980.

[130] Thompson, C.D., "Fourier Transforms in VLSI," *IEEE Transactions on Computers,* Vol. C-32, No. 11, November 1983, pp. 1047-1057.

[131] Thompson, C.D., "The VLSI Complexity of Sorting," *IEEE Transactions on Computers,* Vol. C-32, No. 12, December 1983, pp. 1171-1184.

[132] Toueg, Sam and Ullman, Jeffrey D., "Deadlock-Free Packet Switching Networks," *Proceedings, 11th ACM Symposium on the Theory of Computing,* 1979, pp. 89-98.

[133] Toueg, Sam, "Deadlock- and Livelock-Free Packet Switching Networks," *Proceedings, 12th ACM Symposium on the Theory of Computing,* 1980, pp. 94-99.

[134] Trotter, D., *MOSIS Scalable CMOS Rules,* Version 1.2, 1985.

[135] Ullman, Jeffrey D., *Principles of Database Systems,* Computer Science Press, 1982.

[136] Warshall, S., "A Theorem on Boolean Matrices," *JACM,* Vol. 9, No. 1, January 1962, pp. 11-12.

[137] Wulf, W. and Bell, C.G., "C.mmp - A Multi-Mini-Processor," *Proceedings, AFIPS FJCC,* Vol. 41, Pt. 2, 1972, pp. 765-777.

[138] Xerox Learning Research Group, "The Smalltalk-80 System," *BYTE,* Vol. 6, No. 8, August 1981, pp. 36-48.